Seven Science Articles
on Nanotechnology, Chaos Theory, MATLAB, Solving Equations, Golden Ratio, etc.

Peter I. Kattan

Petra Books
www.PetraBooks.com

Peter I. Kattan, PhD

Correspondence about this book may be sent to the author at one of the following two email addresses:

pkattan@petrabooks.com

info@PetraBooks.com

Seven Science Articles on Nanotechnology, Chaos Theory, MATLAB, Solving Equations, Golden Ratio, etc., written by Peter I. Kattan.
ISBN-13: 979-8-8691-7401-7

All rights reserved. No part of this book may be copied or reproduced without written permission of the author or publisher.
© 2024 Peter I. Kattan

In Loving Memory of My Parents

Seven Science Articles on Nanotechnology, Chaos Theory, MATLAB, Solving Equations, Golden Ratio, etc.

Preface

This book includes seven articles written by the author in various scientific areas. The first two articles were uploaded by the author to the web a few years ago and were mainly available only in electronic form. The other five articles were published as separate chapters in various books written by the author and have an emphasis on using MATLAB[1]. In addition to MATLAB, use is made of the MATLAB Symbolic Math Toolbox[2] in the last five articles.

The first article is a popular science article on nanoscience and nanotechnology. This article was written by the author and uploaded online in electronic form in 2011. In this article, the ratio of surface area to volume is calculated in detail for various shapes, and certain conclusions are made.

The second article is also a popular science article on chaos theory. This article was written by the author and uploaded online in electronic form in 2012. The article includes a simple explanation of chaos theory based on using solely linear algebraic equations. The article has been simplified to such a degree such that it can be widely read and understood. It is noted that the first two articles appear in this book in print for the first time.

The third article shows how to solve algebraic equations using MATLAB. This article first appeared as a chapter in the book "MATLAB for Beginners: A Gentle Approach" which was written by the author and self-published under the Createspace Self-Publishing Platform in 2009. In this article, it is shown how the power and simplicity of MATLAB is used to solve various types of algebraic equations, including linear

---

[1] MATLAB is a registered trademark of the MathWorks, Inc.
[2] The MATLAB Symbolic Math Toolbox is a registered trademark of the MathWorks, Inc.

equations, nonlinear equations, systems of linear simultaneous equations, and systems of nonlinear simultaneous equation.

The fourth article is a short course on solving ordinary differential equations using MATLAB. This article first appeared as a chapter in the book "MATLAB for Beginners: A Gentle Approach – Second Edition" which was written by the author and self-published in 2016 under the Createspace Self-Publishing Platform. In this article emphasis is placed on obtaining analytical closed-form solutions to ordinary differential equations using the MATLAB Symbolic Math Toolbox. Various types of ordinary differential equations are illustrated including first-order ordinary differential equations, second-order ordinary differential equations, and systems of first-order ordinary differential equations.

The fifth article is on Fibonacci Numbers. Specifically, the article delves deeply into this topic while emphasizing using MATLAB for computing these elegant numbers. This article first appeared as a chapter in the book "MATLAB Guide to Fibonacci Numbers and the Golden Ratio" written by the author and published in 2011.

The sixth article is on the Golden Ratio. This article is somehow a continuation of the previous article on Fibonacci Numbers. It is seen that the Golden Ratio is closely linked with the sequence of Fibonacci Numbers. Again, this article delves deeply into calculating the Golden Ratio using MATLAB. The second part of this article examines certain properties of the Golden Ratio. This article first appeared as two chapters in the book "MATLAB Guide to Fibonacci Numbers and the Golden Ratio" written and published by the author in 2011.

The seventh article is on performing regression analysis using MATLAB. Regression analysis is another term for curve fitting. This is a statistical technique used usually in making models of given data from observations. This article first appeared in the book "MATLAB for Beginners: A Gentle Approach – Second Edition" in 2016.

I would like to thank my family members for their help and continued support without which this book would not have been possible. In this edition, I am providing two email addresses for my readers to contact me – pkattan@petrabooks.com and info@PetraBooks.com.

February 2024   Peter I. Kattan

Contents

Seven Science Articles on Nanotechnology, Chaos Theory, MATLAB, Solving Equations, Golden Ratio, etc.

1. Ratio of Surface Area to Volume in Nanotechnology and Nanoscience

This is a concise article detailing the derivations of precise and exact equations for the ratio of surface area to volume for certain geometrical shapes. This ratio is very important in nanomaterials as the noticed increase in this ratio provides these types of materials with distinct properties from their bulk counterparts.

The ratio of surface area to volume (called SAVR in this work) plays a very important role in nanotechnology and nanoscience. This ratio is calculated here for various simple geometrical shapes including the cube, the sphere, the cylinder, the cone, four types of prisms, four types of pyramids, the tetrahedron, the octahedron, the dodecahedron, and the icosahedron.

In addition to calculating SAVR, the article provides also numerical examples of this calculation showing the increase in this ratio for the cube, the sphere, the cylinder, and the cone. This calculation is enhanced with various tables and graphs for SAVR.

The article is the first in our new series on basic nanomechanics. The reader is expected to know the basic mathematical concepts and equations taught in beginning courses of algebra and geometry.

What is Nano?

According to the dictionary.com website, nano is a combining form with the meaning "very small, minute," used in the formation of compound words (*nanoplankton*); in the names of units of measure it has the specific sense "one billionth" (10^{-9}): examples are *nanomole* and *nanosecond*. The origin of the word nano comes from the Greek *nanos* or *nannos* meaning dwarf.

Thus, nano means relating to the very small or extremely small. For example, nanotechnology is technology of the very small. Also, nanoscience means science of the very small. But how small do we go?

Nano is a prefix used just like micro and milli. For example we know that a micrometer is one millionth of a meter. Similarly, a nanometer is one billionth of a meter. This means that one meter is equivalent to one billion nanometers. So a nanometer is a unit of measuring small lengths at the nanoscale.

Another example is a nanosecond which is one billionth of a second. A nanosecond is used to measure extremely small times. Thus, the nanoscale is extremely small. We cannot see nano-objects with our own eyes. We need a powerful microscope in order to see them.

Particles at the nanoscale are called nanoparticles. This is the scale of atoms. Thus, in order to measure atomic distances, we need to use nanometers. The abbreviation for nanometer is nm.

It all started on December 29, 1959 when Richard Feynman gave a lecture at the Annual Meeting of the American Physical Society entitled "There's Plenty of Room at the

Bottom." The full text of the lecture can be found online at the following link:

http://www.zyvex.com/nanotech/feynman.html

In this lecture, Feynman said that he wanted to talk about manipulating and controlling things on a small scale. He said specifically *"Why cannot we write the entire 24 volumes of the Encyclopedia Britannica on the head of a pin?"* Let us now look at the following table which gives the exact definitions of several prefixes for measuring length including nano.

It can be seen in Table 1 that a picometer is 10^{-12} meter. This is a scale even much smaller than the nanoscale. Let us call it the picoscale. In the future, scientists will begin to manipulate objects at the picoscale. Then, a new science will emerge. It will be called picotechnology and picoscience.

Table 1: Powers of 10

Prefix	Power of 10	Exact Factor	Example
tera	10^{12}	1000000000000	1 terabyte = 10^{12} bytes
giga	10^{9}	1000000000	1 gigabyte = 10^{9} bytes
mega	10^{6}	1000000	1 megabyte = 10^{6} bytes
kilo	10^{3}	1000	1 kilometer = 1000 meters
hecto	10^{2}	100	1 hectometer = 100 meters
deca	10^{1}	10	1 decameter = 10 meters
---	10^{0}	1	1 meter = 1 meter
deci	10^{-1}	0.1	1 decimeter = 0.1 meter
centi	10^{-2}	0.01	1 centimeter = 0.01 meter
milli	10^{-3}	0.001	1 millimeter = 0.001 meter
micro	10^{-6}	0.000001	1 micrometer = 10^{-6} meter
nano	10^{-9}	0.000000001	1 nanometer = 10^{-9} meter
pico	10^{-12}	0.000000000001	1 picometer = 10^{-12} meter

Surface Area and Volume

In this article we investigate what happens to the surface area and volume when sizes of objects become small. Surface area is the measure of how much exposed area a solid object has, expressed in square units.

One of the main features of materials at the nanoscale is the increase in their surface area. Actually, what is important is the ratio of surface area to volume. This ratio is called the Surface Area to Volume Ratio and is abbreviated as SAVR. This ratio is the amount of surface are per unit volume of an object. SAVR is measured in units of inverse distance.

We will calculate the values of SAVR for different geometrical objects and investigate what happens to this ratio as sizes become small. We will do this for several geometrical objects, namely the cube, the sphere, the cylinder, the cone, and others.

The Cube

Consider a cube with a side length of 10[1]. The volume of this cube is 10x10x10 = 1000. The area of each side is 10x10 = 100 and since the cube has six sides, the surface area is 6x100 = 600. To calculate the value of SAVR, we divide the surface area by the volume. Thus we get 600/1000 = 0.6.

Next, we repeat the above procedure for a cube with a side length of 5. In this case, the volume is 5x5x5 = 125. The surface area is 6x(5x5) = 150. Thus SAVR = 150/125 = 1.2. Thus, it is clear that when the size of the cube is cut in half (from 10 to 5), the value of SAVR doubles (from 0.6 to 1.2).

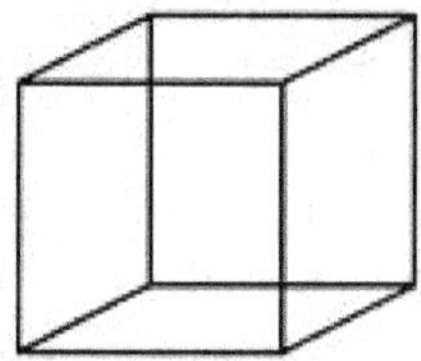

Next, we repeat the above procedure for a cube with a side length of 1. In this case, the volume is 1x1x1 = 1. The surface area is 6x(1x1) = 6. Thus, the value of SAVR is 6/1 =6. Thus, with the decrease in size of the cube from 5 to 1, the value of SAVR increases from 1.2 to 6. Thus, clearly with very small sizes, the surface are to volume ratio increases. This is one of the main features of materials at the nanoscale.

The following table illustrates the values of SAVR for a cube whose side length decreases from 10 to 1 in increments of 1.

Table 2: Calculation of SAVR for a Cube

Side Length of Cube	Volume	Surface Area	SAVR
10	1000	600	0.6
9	729	486	0.667
8	512	384	0.75
7	343	294	0.857
6	216	216	1.0
5	125	150	1.2
4	64	96	1.5
3	27	54	2.0
2	8	24	3.0
1	1	6	6.0

Next, it is better to present the above information in a graph. To do this, let us derive the value of SAVR for an arbitrary cube of side length equal to s. In this case, the volume of the cube is s^3. Since area of each side is s^2, the surface area of the cube is $6s^2$. Thus, we obtain the general formula for SAVR for a cube when we divide the surface area by the volume. Thus dividing $6s^2/s^3 = 6/s$. Thus SAVR $= 6/s$ for a cube of side length s. See Figure 1 for a plot of the side of the cube versus SAVR.

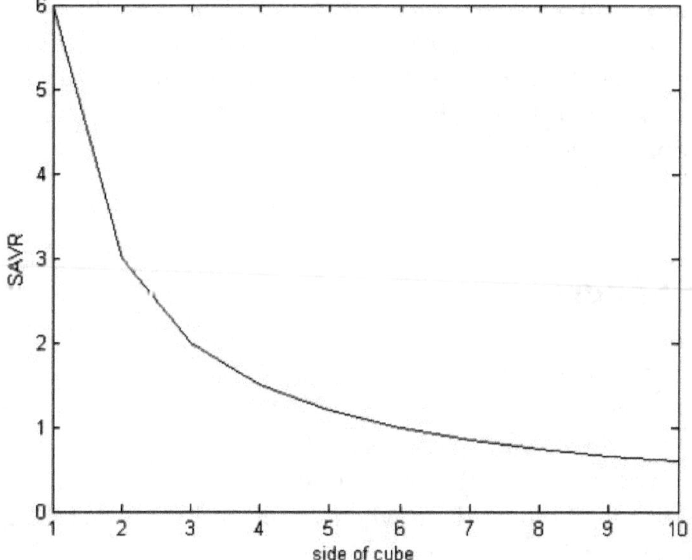

Figure 1: SAVR Variation for a Cube

It is clear from the formula SAVR = 6/s that when s becomes very small, SAVR becomes very large. As s approaches zero, the value of SAVR approaches infinity. Thus for an infinitesimally small cube, the value of SAVR becomes infinitely large. Let us look at some numerical examples of cubes at the nanoscale.

$$\text{SAVR} = \frac{6}{s}$$

For a cube of side length 100 nm, let us calculate the value of SAVR. Since s = 100 nm = 100 x 10^{-9} m = 1 x 10^{-7} m, then we obtain SAVR = 6/(1 x 10^{-7}) m^{-1} = 6 x 10^{7} m^{-1}. This value is 60,000,000 which is extremely large.

Let us see what happens to the value of SAVR when the side length of the cube is reduced to 10 nm. Thus, s = 10 nm

$= 10 \times 10^{-9}$ m $= 1 \times 10^{-8}$ m. Thus, SAVR $= 6/(1 \times 10^{-8})$ m^{-1} $= 6 \times 10^{8}$ m^{-1}. This value is 600,000,000 which is a lot larger than the previous value. It is clear from these two examples that we obtain extremely large values of the surface are to volume ratio for cubes at the nanoscale.

The Sphere

Consider a sphere of radius equal to $10^{[2]}$. The volume of this sphere is 4xπx10x10x10/3 = 4188.79, where π is the well known constant of the circle (the ratio of the area of a circle to its perimeter). The surface area of this sphere is 4xπx10x10 = 1256.64. Thus when we divide 1256.64/4188.29, we obtain a value of SAVR = 0.30.

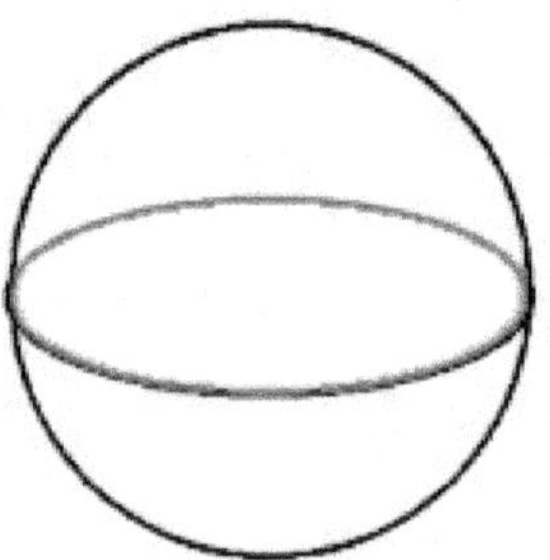

Let us repeat the above procedure for a sphere of radius equal to 5. The volume of this sphere is 4xπ5x5x5/3 = 523.60. The surface area of this sphere is 4xπx5x5 = 314.16. Thus we obtain the value of SAVR = 314.16/523.60 = 0.6. Thus, clearly the value of SAVR doubles when the radius of the sphere is reduced from 10 to 5.

Next, we repeat the above procedure for a sphere of radius equal to 1. The volume of this sphere is 4xπx1x1x1/3 = 4.189. The surface area of this sphere is 4xπx1x1 = 12.566. In this case, the value of SAVR = 12.566/4.189 = 3.0. This the value of the surface area to volume ratio increase substantially as the radius of the sphere is reduced.

Table 3: Calculation of SAVR for a Sphere

Radius of Sphere	Volume	Surface Area	SAVR
10	4188.79	1256.64	0.3
9	3053.63	1017.88	0.333
8	2144.66	804.25	0.375
7	1436.75	615.75	0.429
6	904.28	452.39	0.5
5	523.60	314.16	0.6
4	268.08	201.06	0.75
3	113.10	113.10	1.0
2	33.51	50.27	1.5
1	4.188	12.566	3.0

Table 2 illustrates the values of SAVR for a sphere whose radius decreases from 10 to 1 in increments of 1.

Next, it is better to present the above information in a graph. To do this, let us derive the value of SAVR for an arbitrary sphere of radius equal to r. In this case, the volume of the sphere is $4x\pi xr^3/3$. The surface area of the sphere is $4x\pi xr^2$. Thus, we obtain the general formula for SAVR for a sphere when we divide the surface area by the volume. Thus dividing $4x\pi xr^2/(4x\pi xr^3/3) = 3/r$. Thus SAVR = 3/r for a sphere of radius r. See Figure 2 for the variation of SAVR versus the radius of the sphere.

$$SAVR = \frac{3}{r}$$

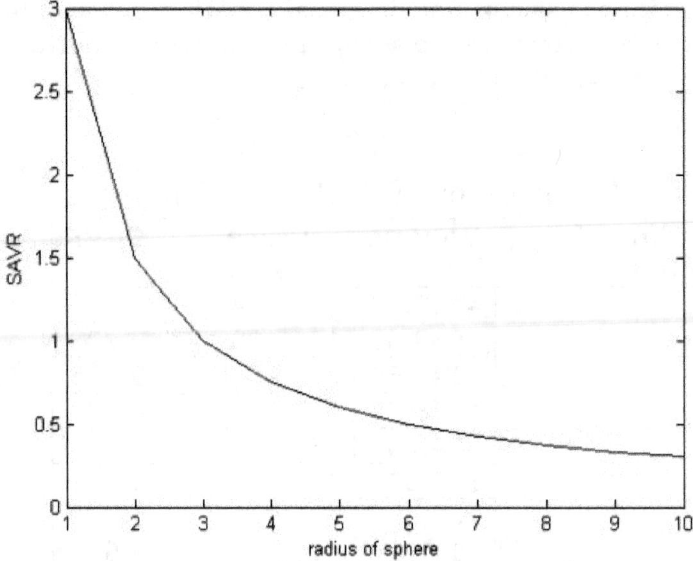

Figure 2: Variation of SAVR for a Sphere

It is clear from the formula SAVR = 3/r that when r becomes very small, SAVR becomes very large. As r approaches zero, the value of SAVR approaches infinity. Thus for an infinitesimally small sphere, the value of SAVR becomes infinitely large. Let us look at some numerical examples of spheres at the nanoscale.

For a sphere of radius 100 nm, let us calculate the value of SAVR. Since r = 100 nm = 100 x 10^{-9} m = 1 x 10^{-7} m, then we obtain SAVR = 3/(1 x 10^{-7}) m^{-1} = 3 x 10^{7} m^{-1}. This value is 30,000,000 which is extremely large.

Let us see what happens to the value of SAVR when the radius of the sphere is reduced to 10 nm. Thus, r = 10 nm = 10 x 10^{-9} m = 1 x 10^{-8} m. Thus, SAVR = 3/(1 x 10^{-8}) m^{-1} = 3 x 10^{8} m^{-1}. This value is 300,000,000 which is a lot larger than the previous value. It is clear from these two examples that we obtain extremely large values of the surface are to volume ratio for spheres at the nanoscale.

The Cylinder

Consider a cylinder of radius 10[3] and height 100. The volume of this cylinder is πx10x10x100 = 31415.93. The surface area of the cylinder is calculated by adding the two circular areas with the circumferential area. This is obtained for this cylinder as follows. Surface area = 2xπx10x10 + 2xπx10x100 = 6911.50. Thus SAVR = 6911.50/31415.93 = 0.22.

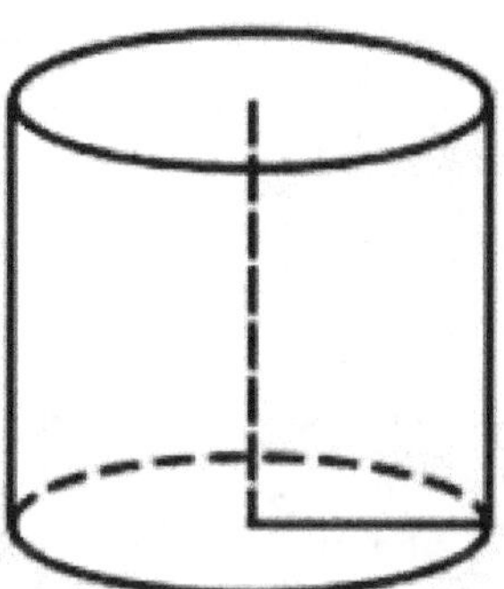

Let us now repeat the above procedure for a cylinder of radius 1 and height 10. The volume of this cylinder is πx1x1x10 = 31.416. The surface area is 2xπx1x1 + 2xπx1x10 = 69.115. Thus, SAVR = 69.115/31.416 = 2.2. Thus there is a substantial increase in the value of SAVR when the radius and height of the cylinder are both reduced. Clearly this is one of the characteristics of materials at the nanoscale.

Let us now derive the value of SAVR for an arbitrary cylinder of radius r and height h. The volume of this cylinder is $\pi r^2 h$. The surface area of this cylinder is $2\pi r^2 + 2\pi rh$. Thus,

SAVR = $(2\pi r^2 + 2\pi rh)/ \pi r^2 h = 2/r + 2/h$. This simple formula is used in generating the following table of SAVR values for a cylinder with radius and height that decrease in size from 10 and 100 to 1 and 10, respectively.

Table 4: Calculation of SAVR for a Cylinder

Radius of Cylinder	Height of Cylinder	SAVR
10	100	0.22
9	90	0.244
8	80	0.275
7	70	0.314
6	60	0.367
5	50	0.44
4	40	0.55
3	30	0.733
2	20	1.1
1	10	2.2

Since SAVR = $2/r + 2/h$, it is very clear that as both r and h go to zero, the value of SAVR goes to infinity. Thus, as the radius and height of a cylinder become smaller, the surface area to volume ratio becomes larger.

$$SAVR = \frac{2}{r} + \frac{2}{h}$$

Let us investigate a numerical example of a cylinder at the nanoscale. Consider a cylinder of radius equal to 10 nm and height equal to 100 nm. Thus, r = 10 nm = 10×10^{-9} = 1×10^{-8} m and h = 100 nm = 100×10^{-9} m = 1×10^{-7} m. Thus, we get SAVR = $2/(1 \times 10^{-8}) + 2/(1 \times 10^{-7})$ m^{-1} = 2.2×10^8 m^{-1} which is 220,000,000 – clearly a very large number. Thus, we get huge values for SAVR for cylinders at the nanoscale.

The Cone

Consider a cone of radius 10[4] and height 100. The volume of this cone is $\pi\times10\times10\times100/3 = 10471.98$. The surface area of the cone is calculated by adding the circular area at the bottom with the circumferential area. This is obtained for this cone as follows. Surface area $= \pi\times10\times10 + \pi\times10\times\sqrt{10^2 + 100^2} = 3471.421$. Thus SAVR $= 3471.421/10471.98 = 0.331$.

Let us now repeat the above procedure for a cone of radius 1 and height 10. The volume of this cone is $\pi\times1\times1\times10/3 = 10.472$. The surface area is $\pi\times1\times1 + \pi\times1\times\sqrt{1^2 + 10^2} = 34.714$. Thus, SAVR $= 34.714/10.472 = 3.31$. Thus there is a substantial increase in the value of SAVR when the radius and height of the cone are both reduced. Clearly this is one of the characteristics of materials at the nanoscale.

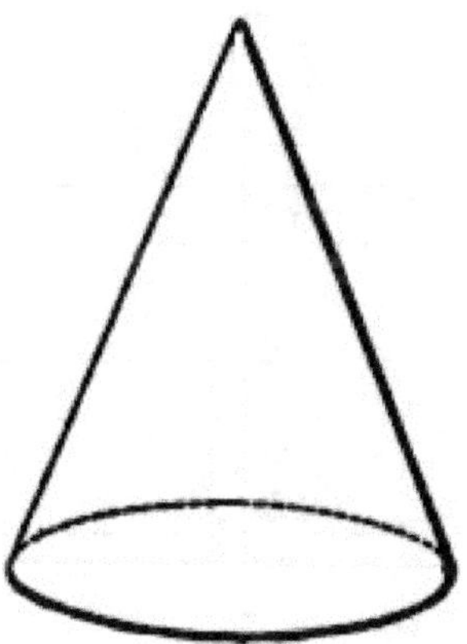

Let us now derive the value of SAVR for an arbitrary cone of radius r and height h. The volume of this cylinder is $(1/3)\pi r^2 h$. The surface area of this cone is $\pi r^2 + \pi r\sqrt{r^2 + h^2}$.

Check the Appendix for details how we got these formulas for volume and surface area of a cone. Thus, $\text{SAVR} = 3(\pi r^2 + \pi r \sqrt{r^2 + h^2})/ \pi r^2 h = 3/h + 3\sqrt{\dfrac{1}{h^2} + \dfrac{1}{r^2}}$. This simple formula is used in generating the following table of SAVR values for a cone with radius and height that decrease in size from 10 and 100 to 1 and 10, respectively.

Since $\text{SAVR} = 3/h + 3\sqrt{\dfrac{1}{h^2} + \dfrac{1}{r^2}}$, it is very clear that as both r and h go to zero, the value of SAVR goes to infinity. Thus, as the radius and height of a cone become smaller, the surface area to volume ratio becomes larger.

Table 5: Calculation of SAVR for a Cone

Radius of Cylinder	Height of Cylinder	SAVR
10	100	0.331
9	90	0.368
8	80	0.414
7	70	0.474
6	60	0.552
5	50	0.663
4	40	0.829
3	30	1.11
2	20	1.66
1	10	3.31

$$\text{SAVR} = \frac{3}{h} + 3\sqrt{\frac{1}{h^2} + \frac{1}{r^2}}$$

Let us investigate a numerical example of a cone at the nanoscale. Consider a cone of radius equal to 10 nm and height equal to 100 nm. Thus, r = 10 nm = 10 x 10^{-9} = 1 x 10^{-8} m and h = 100 nm = 100 x 10^{-9} m = 1 x 10^{-7} m. Thus, we get SAVR

$$= 3/(1 \times 10^{-7}) + 3\sqrt{\frac{1}{\left(3x10^{-7}\right)^2} + \frac{1}{\left(3x10^{-8}\right)^2}}\ m^{-1} = 1.30 \times 10^{8}\ m^{-1}$$

which is 130,000,000 – clearly a very large number. Thus, we get huge values for SAVR for cones at the nanoscale.

The Triangular Prism

Let us now derive the value of SAVR for an arbitrary prism of height h and a triangular base of sides a, b, and c. The volume of this prism is (1/2) x.ah where x is the perpendicular line from one vertex of the base to the opposite side (side a in this case). To simplify the calculations, let us consider the triangular base to be equilateral, that is all three sides have the same length. Let us call this length s. Thus, we have a+b+c = 3s. In this special case, the perpendicular length x =

$$\sqrt{s^2 - \left(\frac{s}{2}\right)^2} = \frac{\sqrt{3}}{2}s$$

Thus, the volume of this prism is equal to $\frac{1}{2}\frac{\sqrt{3}}{2}sah$. The surface area of this prism is equal to 2(1/2 x.a) + (a+b+c)h which is equal to $\frac{\sqrt{3}}{2}sa + 3sh$. Check the Appendix for details how we got these formulas for volume and surface area of a triangular prism.

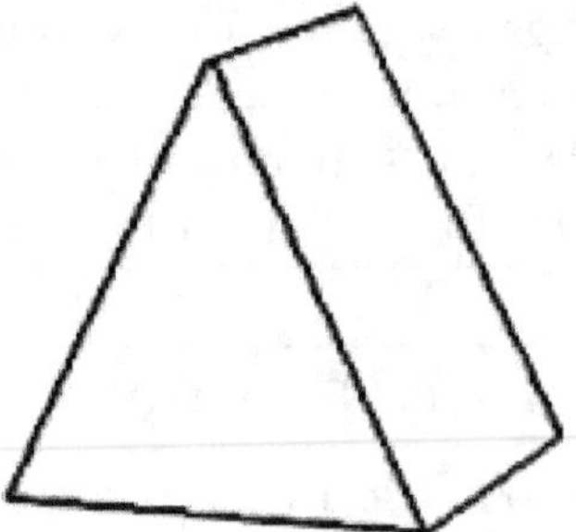

$$\frac{\frac{\sqrt{3}}{2}sa+3s\,h}{\frac{\sqrt{3}}{4}s\,ah}=\frac{2}{h}+\frac{4\sqrt{3}}{a}$$

Thus, SAVR = (above) , but a = s in this special case of an equilateral triangular base, thus, we get the following simplified formula for a prism with an equilateral triangular base:

$$SAVR = \frac{2}{h}+\frac{4\sqrt{3}}{s}$$

It is very clear from the above equation that as both s and h go to zero, the value of SAVR goes to infinity. Thus, as the height and side of a prism become smaller, the surface area to volume ratio becomes larger.

The Rectangular Prism

Let us now derive the value of SAVR for an arbitrary prism of height h and a rectangular base of sides a and b. The volume of this prism is abh. The surface area of this prism is equal to 2ab + 2ah + 2bh. Check the Appendix for details how we got these formulas for volume and surface area of a rectangular prism.

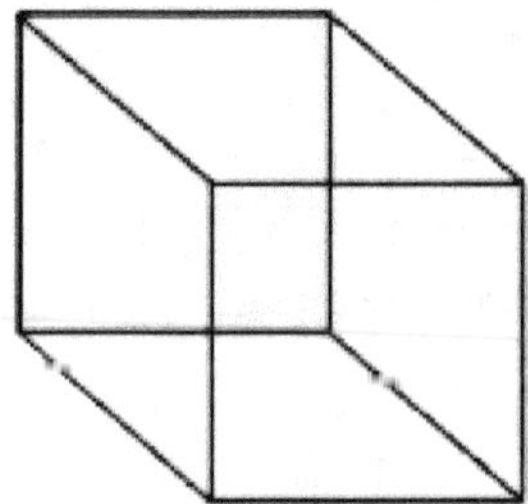

Thus, SAVR $= \dfrac{2ab + 2ah + 2bh}{abh} = \dfrac{2}{h} + \dfrac{2}{b} + \dfrac{2}{a}$, thus, we get the following simplified formula for a rectangular prism:

$$\text{SAVR} = \frac{2}{h} + \frac{2}{a} + \frac{2}{b}$$

It is very clear from the above equation that as a, b, and h go to zero, the value of SAVR goes to infinity. Thus, as sides of a prism become smaller, the surface area to volume ratio becomes larger.

The Pentagonal Prism

Let us now derive the value of SAVR for an arbitrary prism of height h and a base in the shape of a regular pentagon of side s. The volume of this prism is ½ (x.s)(5)h = (5/2) s.x.h, where x is the perpendicular distance from the center of the base to one of the opposite sides given by x $=$

$$\sqrt{s^2 - \left(\frac{s}{2}\right)^2} = \frac{\sqrt{3}}{2}s \, .$$

Thus, the volume of this prism is equal to

$$\frac{5}{2}s\frac{\sqrt{3}}{2}sh = \frac{5\sqrt{3}}{4}s^2h$$

. The surface area of this prism is equal to 5sh + 2(1/2 x.s)(5) = 5sh + 5x.s = $5sh + \frac{5\sqrt{3}}{2}s^2$. Check the Appendix for details how we got these formulas for volume and surface area of a pentagonal prism.

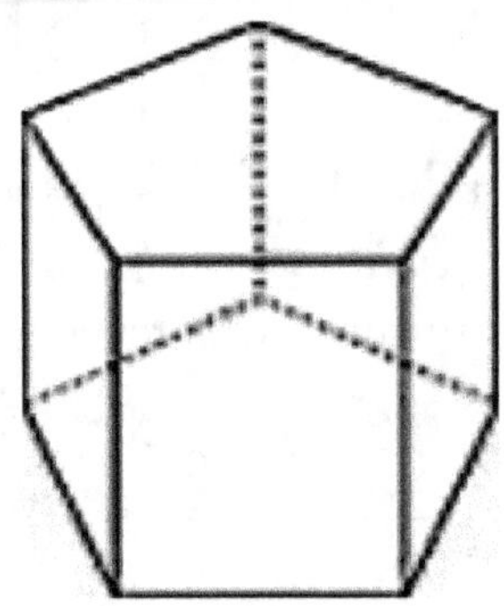

$$\frac{5sh + \frac{5\sqrt{3}}{2}s^2}{\frac{5\sqrt{3}}{4}s^2h} = \frac{4\sqrt{3}}{3s} + \frac{2}{h}$$

Thus, SAVR = , thus, we get the following simplified formula for a pentagonal prism:

$$\text{SAVR} = \frac{2}{h} + \frac{4\sqrt{3}}{3s}$$

It is very clear from the above equation that as both s and h go to zero, the value of SAVR goes to infinity. Thus, as the height and side of a prism become smaller, the surface area to volume ratio becomes larger.

The Hexagonal Prism

Let us now derive the value of SAVR for an arbitrary prism of height h and a base in the shape of a regular hexagon of side s. The volume of this prism is ½ (x.s)(6)h = 3s.x.h

where x is the perpendicular distance from the center of the base to one of the opposite sides given by $x = \sqrt{s^2 - \left(\dfrac{s}{2}\right)^2} = \dfrac{\sqrt{3}}{2}s$.

Thus, the volume of this prism is equal to $3s\dfrac{\sqrt{3}}{2}sh = \dfrac{3\sqrt{3}}{2}s^2h$. The surface area of this prism is equal to 6sh + 2(1/2 x.s)(6) = 6sh + 6x.s = $6sh + 3\sqrt{3}s^2$. Check the Appendix for details how we got these formulas for volume and surface area of a hexagonal prism.

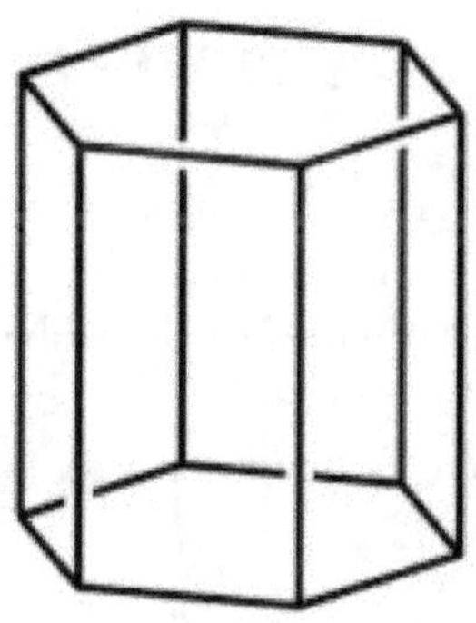

$$\frac{6sh + 3\sqrt{3}s^2}{\dfrac{3\sqrt{3}}{2}s^2 h} = \frac{4\sqrt{3}}{3s} + \frac{2}{h}$$

Thus, SAVR = , thus, we get the following simplified formula for a hexagonal prism:

$$SAVR = \frac{2}{h} + \frac{4\sqrt{3}}{3s}$$

It is very clear from the above equation that as both s and h go to zero, the value of SAVR goes to infinity. Thus, as the height and side of a prism become smaller, the surface area to volume ratio becomes larger.

Note that the above formula is similar to the one derived for a pentagonal prism. This formula of SAVR seems to apply to any prism with a base in the shape of a regular polygon. We will prove this in the next section.

The General n-Polygonal Prism

Let us now derive the value of SAVR for an arbitrary prism of height h and a base in the shape of a regular polygon of side s. The number of sides of the polygon is n. The volume of this prism is ½ (x.s)(n)h = 3s.x.h where x is the perpendicular distance from the center of the base to one of the opposite sides given by $x = \sqrt{s^2 - \left(\dfrac{s}{2}\right)^2} = \dfrac{\sqrt{3}}{2}s$.

Thus, the volume of this prism is equal to $\dfrac{\sqrt{3}}{4}snh = \dfrac{\sqrt{3}}{4}s^2nh$.

The surface area of this prism is equal to nsh + 2(1/2 x.s)(n) =

nsh + nx.s = $nsh + \dfrac{\sqrt{3}}{2}s^2n$.

Thus, SAVR = $\dfrac{nsh + \dfrac{\sqrt{3}}{2}s^2n}{\dfrac{\sqrt{3}}{4}s^2nh} = \dfrac{4\sqrt{3}}{3s} + \dfrac{2}{h}$, thus, we get the following simplified formula for a polygonal prism:

$$SAVR = \dfrac{2}{h} + \dfrac{4\sqrt{3}}{3s}$$

It is very clear from the above equation that as both s and h go to zero, the value of SAVR goes to infinity. Thus, as the height and side of a prism become smaller, the surface area to volume ratio becomes larger. The above formula is valid for any general n-polygonal prism irrespective of the number of sides of the base. The above is the exactly the same formula derived for both the pentagonal prism and the hexagonal prism.

The Triangular Pyramid

Let us now derive the value of SAVR for an arbitrary pyramid of height h and a triangular base of equilateral sides s. The volume of this pyramid is (1/3) (1/2 x.s)h = (1/6) x.sh where the perpendicular length x = $\sqrt{s^2 - \left(\dfrac{s}{2}\right)^2} = \dfrac{\sqrt{3}}{2}s$, and the inclined length l = $\sqrt{x^2 + h^2}$. Here, x is the perpendicular from one vertex of the base to the opposite side. Thus, the volume of this pyramid is equal to $\dfrac{1}{6}\dfrac{\sqrt{3}}{2}s^2h$. The surface area of this pyramid $3(1/2\ s.l) + \tfrac{1}{2}\ s.x$ which is equal to $\dfrac{1}{2}\dfrac{\sqrt{3}}{2}s^2 + \dfrac{3}{2}s\sqrt{\dfrac{3}{4}s^2 + h^2}$. Check the Appendix for details how we got these formulas for volume and surface area of a triangular pyramid.

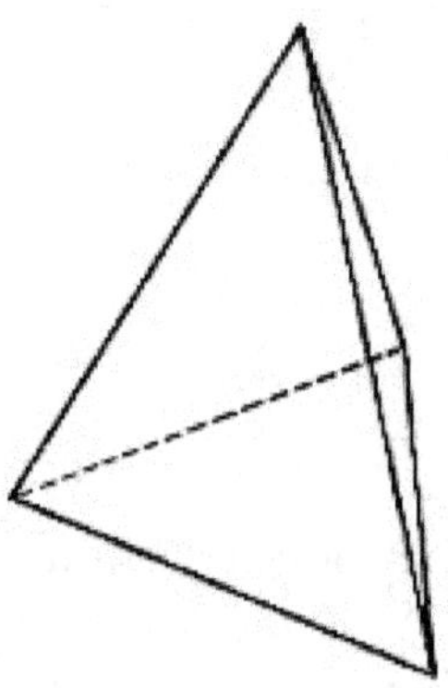

Thus,

$$\text{SAVR} = \dfrac{\dfrac{\sqrt{3}}{4}s^2 + \dfrac{3}{2}s\sqrt{\dfrac{3}{4}s^2 + h^2}}{\dfrac{\sqrt{3}}{12}s^2h} = \dfrac{3}{h} + 6\sqrt{3}\sqrt{\dfrac{3}{4h^2} + \dfrac{1}{s^2}}$$

, thus, we get the following simplified formula for a pyramid with an equilateral triangular base:

$$\text{SAVR} = \frac{3}{h} + 6\sqrt{3}\sqrt{\frac{3}{4h^2} + \frac{1}{s^2}}$$

It is very clear from the above equation that as both s and h go to zero, the value of SAVR goes to infinity. Thus, as the height and side of a pyramid become smaller, the surface area to volume ratio becomes larger.

The Square Pyramid

Let us now derive the value of SAVR for an arbitrary pyramid of height h and a square base of side s. The volume of this pyramid is $(1/3)\ s^2h$. The surface area of this pyramid is $4(1/2\ s.l) + s^2$, where the inclined length $l = \sqrt{x^2 + h^2}$. Check the Appendix for details how we got these formulas for volume and surface area of a square pyramid.

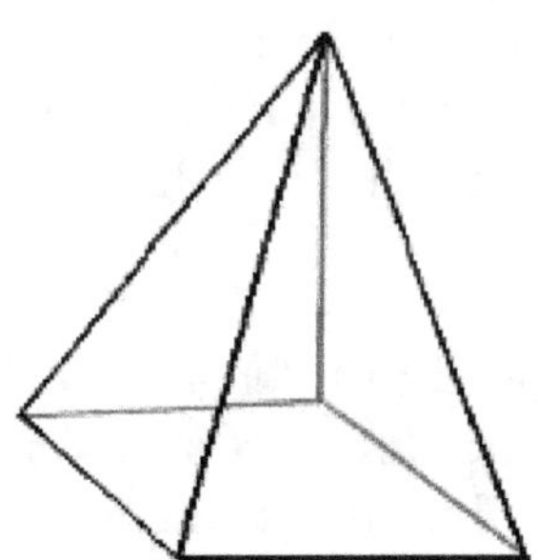

Thus, SAVR $= \dfrac{s^2 + 2s\sqrt{\dfrac{3}{4}s^2 + h^2}}{\dfrac{1}{3}s^2 h} = \dfrac{3}{h} + 6\sqrt{\dfrac{3}{4h^2} + \dfrac{1}{s^2}}$, thus, we get the following simplified formula for a pyramid with a square base:

$$\text{SAVR} = \frac{3}{h} + 6\sqrt{\frac{3}{4h^2} + \frac{1}{s^2}}$$

It is very clear from the above equation that as both s and h go to zero, the value of SAVR goes to infinity. Thus, as the height and side of a pyramid become smaller, the surface area to volume ratio becomes larger.

The Pentagonal Pyramid

Let us now derive the value of SAVR for an arbitrary pyramid of height h and a base in the shape of a regular pentagon of side s. The volume of this pyramid is (1/3)(5)(1/2 x.s)h, where the perpendicular length $x = \sqrt{s^2 - \left(\dfrac{s}{2}\right)^2} = \dfrac{\sqrt{3}}{2}s$. Here, x is the perpendicular line from one vertex of the base to the opposite side. Thus, the volume of this pyramid is equal to $\dfrac{5}{6}\dfrac{\sqrt{3}}{2}s^2 h$. The surface area of this pyramid is ½ (s.l)(5) + ½ (x.s)(5), which is equal to $\dfrac{5}{2}\dfrac{\sqrt{3}}{2}s^2 + \dfrac{5}{2}s\sqrt{\dfrac{3}{4}s^2 + h^2}$, where the inclined length $1 = \sqrt{x^2 + h^2}$. Check the Appendix for details how we got these formulas for volume and surface area of a pentagonal pyramid.

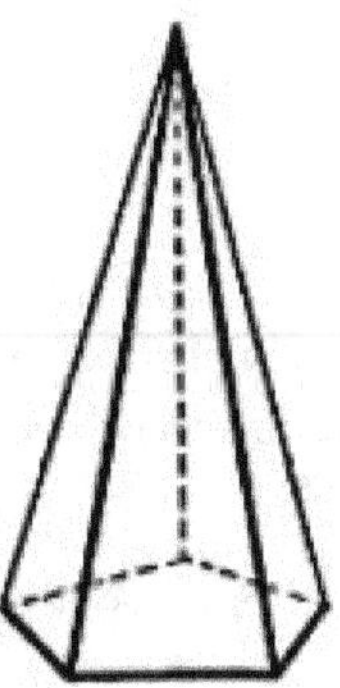

Thus,

$$\frac{\dfrac{5\sqrt{3}}{4}s^2 + \dfrac{5}{2}s\sqrt{\dfrac{3}{4}s^2 + h^2}}{\dfrac{5\sqrt{3}}{12}s^2 h} = \frac{3}{h} + 2\sqrt{3}\sqrt{\dfrac{3}{4h^2} + \dfrac{1}{s^2}}$$

SAVR =

thus, we get the following simplified formula for a pyramid with a base in the shape of a regular pentagon:

$$\text{SAVR} = \frac{3}{h} + 2\sqrt{3}\sqrt{\dfrac{3}{4h^2} + \dfrac{1}{s^2}}$$

It is very clear from the above equation that as both s and h go to zero, the value of SAVR goes to infinity. Thus, as the height and side of a pyramid become smaller, the surface area to volume ratio becomes larger.

The Hexagonal Pyramid

Let us now derive the value of SAVR for an arbitrary pyramid of height h and a base in the shape of a regular hexagon of side s. The volume of this pyramid is $(1/3)(6)(1/2$ x.s)h, where the perpendicular length x $= \sqrt{s^2 - \left(\dfrac{s}{2}\right)^2} = \dfrac{\sqrt{3}}{2}s$. Here, x is the perpendicular line from one vertex of the base to the opposite side. Thus, the volume of this pyramid is equal to $\dfrac{\sqrt{3}}{2}s^2 h$. The surface area of this pyramid is ½ (s.l)(6) + ½ (x.s)(6), which is equal to $3\dfrac{\sqrt{3}}{2}s^2 + 3s\sqrt{\dfrac{3}{4}s^2 + h^2}$, where the inclined length l $= \sqrt{x^2 + h^2}$. Check the Appendix for details how we got these formulas for volume and surface area of a hexagonal pyramid.

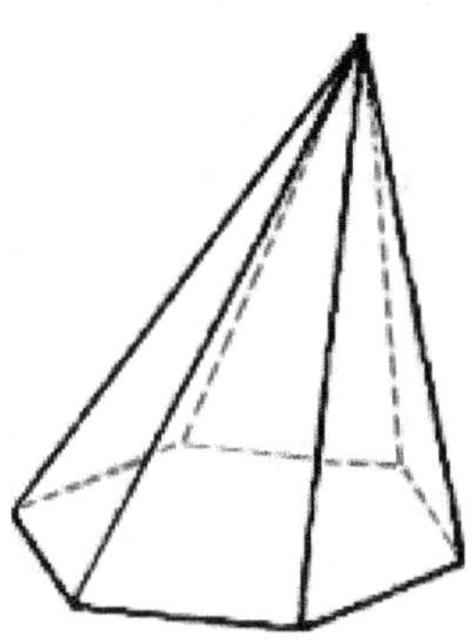

Thus,

$$\text{SAVR} = \frac{\dfrac{3\sqrt{3}}{2}s^2 + 3s\sqrt{\dfrac{3}{4}s^2 + h^2}}{\dfrac{\sqrt{3}}{2}s^2 h} = \frac{3}{h} + 2\sqrt{3}\sqrt{\dfrac{3}{4h^2} + \dfrac{1}{s^2}}$$

, thus, we get the following simplified formula for a pyramid with a base in the shape of a regular hexagon:

$$\text{SAVR} = \frac{3}{h} + 2\sqrt{3}\sqrt{\frac{3}{4h^2} + \frac{1}{s^2}}$$

It is very clear from the above equation that as both s and h go to zero, the value of SAVR goes to infinity. Thus, as the height and side of a pyramid become smaller, the surface area to volume ratio becomes larger.

It should be noted that we obtained a similar formula for the hexagonal pyramid as that obtained previously for the pentagonal pyramid. We will prove this fact in the next section.

The General n-Polygonal Pyramid

Let us now derive the value of SAVR for an arbitrary pyramid of height h and a base in the shape of a regular polygon of side s. The number of side of the polygon is n. The volume of this pyramid is $(1/3)(n)(1/2 \ x.s)h$, where the perpendicular length $x = \sqrt{s^2 - \left(\dfrac{s}{2}\right)^2} = \dfrac{\sqrt{3}}{2}s$. Here, x is the perpendicular line from one vertex of the base to the opposite side. Thus, the volume of this pyramid is equal to $\dfrac{\sqrt{3}}{12}s^2 n h$. The surface area of this pyramid is $\frac{1}{2} \ (s.l)(n) + \frac{1}{2} \ (x.s)(n)$, which is equal to $\dfrac{\dfrac{\sqrt{3}}{4}s^2 n + \dfrac{1}{2}sn\sqrt{\dfrac{3}{4}s^2 + h^2}}{}$, where the inclined length $l = \sqrt{x^2 + h^2}$. Check the

Thus,

$$\text{SAVR} = \frac{\dfrac{\sqrt{3}}{4}s^2 n + \dfrac{1}{2}sn\sqrt{\dfrac{3}{4}s^2 + h^2}}{\dfrac{\sqrt{3}}{12}s^2 nh} = \frac{3}{h} + 2\sqrt{3}\sqrt{\frac{3}{4h^2} + \frac{1}{s^2}},$$

thus, we get the following simplified formula for a pyramid with a base in the shape of a regular polygon:

$$\text{SAVR} = \frac{3}{h} + 2\sqrt{3}\sqrt{\frac{3}{4h^2} + \frac{1}{s^2}}$$

It is very clear from the above equation that as both s and h go to zero, the value of SAVR goes to infinity. Thus, as the height and side of a pyramid become smaller, the surface area to volume ratio becomes larger. It should be noted that the above formula is valid for any general n-polygonal pyramid irrespective of the number of sides of the polygon. It is exactly the same formula derived for the pentagonal pyramid and the hexagonal pyramid.

Comparison

Comparing the formulas derived previously for general n-polygonal prisms and pyramids, it is noticed that the value of SAVR for a pyramid is 1.5 times larger than that for a prism. If you multiply the formula for the general n-polygonal prism by 1.5, we get exactly the formula for the general n-polygonal pyramid. This means that pyramids will provide better materials at the nanoscale than prisms. Thus, materials in the shape of pyramids are recommended to be used as nanomaterials than materials in the shape of prisms. This is primarily because of the higher value of SAVR of pyramids.

The Tetrahedron

The tetrahedron is a solid composed of four identical faces. Each face is an equilateral triangle of side s. The volume of the tetrahedron is $\dfrac{\sqrt{2}}{12}s^3$ and its surface area is $\sqrt{3}s^2$. Thus.

$$SAVR = \dfrac{\sqrt{3}s^2}{\dfrac{\sqrt{2}}{12}s^3} = \dfrac{6\sqrt{6}}{s}.$$

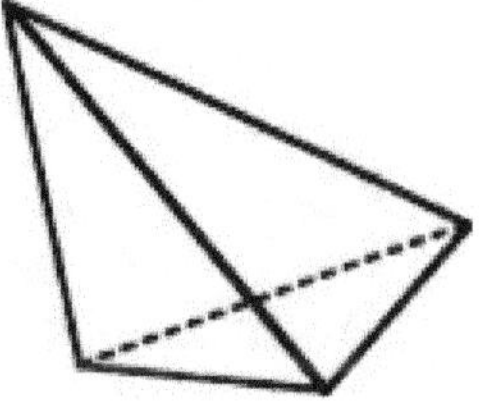

Thus, we get the following simplified formula for a tetrahedron with sides s:

$$SAVR = \dfrac{6\sqrt{6}}{s}$$

It is very clear from the above equation that as s goes to zero, the value of SAVR goes to infinity. Thus, as the side of a tetrahdron becomes smaller, the surface area to volume ratio becomes larger.

Note that the tetrahedron is the same as the triangular pyramid with equal faces.

The Octahedron

The octahedron is a solid composed of eight identical faces. Each face is an equilateral triangle of side s. The volume of the octahedron is $\dfrac{\sqrt{2}}{3}s^3$ and its surface area is $2\sqrt{3}s^2$. Thus.

$$\text{SAVR} = \frac{2\sqrt{3}s^2}{\dfrac{\sqrt{2}}{3}s^3} = \frac{3\sqrt{6}}{s}.$$

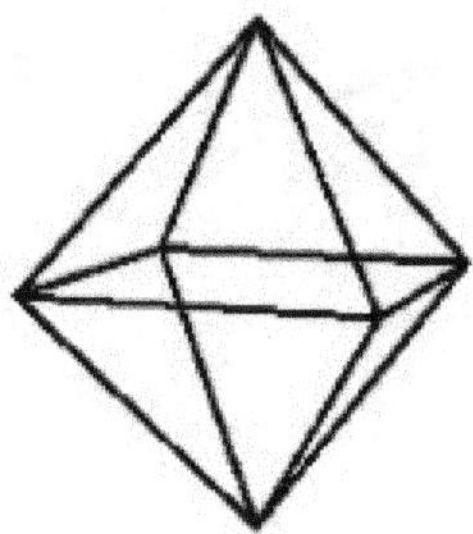

Thus, we get the following simplified formula for a octahedron with sides s:

$$\text{SAVR} = \frac{3\sqrt{6}}{s}$$

It is very clear from the above equation that as s goes to zero, the value of SAVR goes to infinity. Thus, as the side of an octahdron becomes smaller, the surface area to volume ratio becomes larger.

Thus, it is clear that we obtain the same formula for an octahedron as that obtained for a tetrahedron except for a factor of 2.

The Dodecahedron

The dodecahedron is a solid composed of twelve identical faces. Each face is a regular pentagon of side s. The volume of the dodecahedron is $\dfrac{15+7\sqrt{5}}{4}s^{3}$ and its surface area is $3\sqrt{25+10\sqrt{5}}s^{2}$. Thus. SAVR $=$

$$\frac{3\sqrt{25+10\sqrt{5}}s^{2}}{\dfrac{15+7\sqrt{5}}{4}s^{3}}=\frac{12\sqrt{25+10\sqrt{5}}}{\left(15+7\sqrt{5}\right)s}.$$

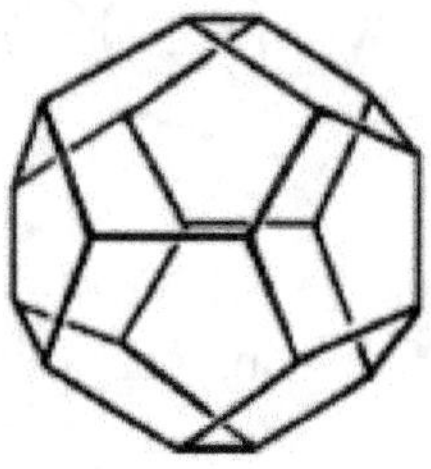

Thus, we get the following simplified formula for a dodecahedron with sides s:

$$SAVR=\frac{12\sqrt{25+10\sqrt{5}}}{\left(15+7\sqrt{5}\right)s}$$

It is very clear from the above equation that as s goes to zero, the value of SAVR goes to infinity. Thus, as the side of an dodecahdron becomes smaller, the surface area to volume ratio becomes larger.

The Icosahedron

The icosahedron is a solid composed of twenty identical faces. Each face is an equilateral triangle of side s. The volume of the icosahedron is $\dfrac{5}{12}\left(3+\sqrt{5}\right)s^3$ and its surface area is $5\sqrt{3}s^2$.

Thus. SAVR $= \dfrac{5\sqrt{3}s^2}{\dfrac{5\left(3+\sqrt{5}\right)}{12}s^3} = \dfrac{12\sqrt{3}}{\left(3+\sqrt{5}\right)s}$

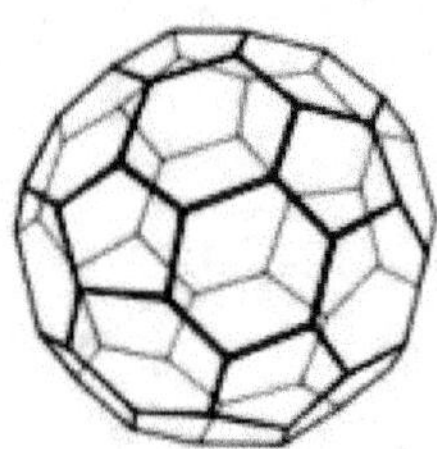

Thus, we get the following simplified formula for an icosahedron with sides s:

$$SAVR = \dfrac{12\sqrt{3}}{\left(3+\sqrt{5}\right)s}$$

It is very clear from the above equation that as s goes to zero, the value of SAVR goes to infinity. Thus, as the side of an icosahdron becomes smaller, the surface area to volume ratio becomes larger.

Comparison

Finally, we compare the values of SAVR that we derived for the tetrahedron, the octahedron, the dodecahedron, and the icosahedron. The following is a summary of the results:

Tetrahedron (4 faces)
$$\text{SAVR} = \frac{6\sqrt{6}}{s} \approx \frac{14.70}{s}$$

Octahedron (8 faces)
$$\text{SAVR} = \frac{3\sqrt{6}}{s} \approx \frac{7.35}{s}$$

Dodecahedron (12 faces)
$$\text{SAVR} = \frac{12\sqrt{25+10\sqrt{5}}}{\left(15+7\sqrt{5}\right)s} \approx \frac{2.69}{s}$$

Icosahedron (20 faces)
$$\text{SAVR} = \frac{12\sqrt{3}}{\left(3+\sqrt{5}\right)s} \approx \frac{3.97}{s}$$

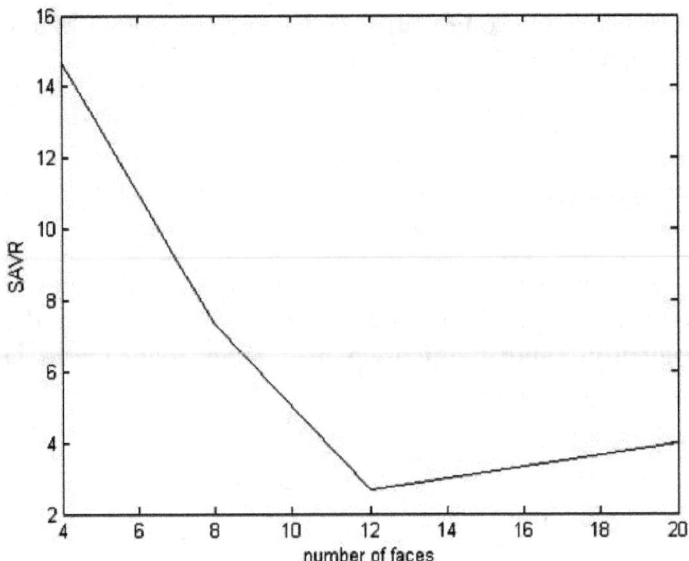

Figure3: SAVR Values vs. Number of Faces

Figure 3 shows a graph comparing the number of faces of the object with the value of SAVR. It is clear from the figure that the optimal number of faces is close to 12 for the lowest value of SAVR. However, it should be noted that the tetrahedron with four faces has the largest value of SAVR. Larger values of SAVR provide for better materials at the nanoscale.

Mathematical Equations Used in This Article

<u>The Cube</u>

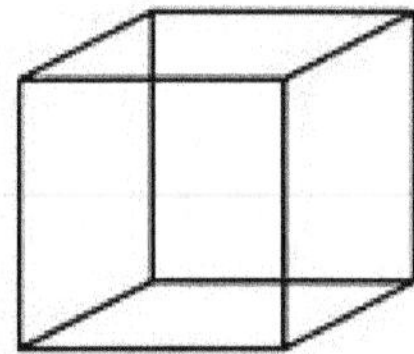

Assume the cube to have a side of length $= s$

Surface area of cube $= 6s^2$

Volume of cube $= s^3$

<u>The Sphere</u>

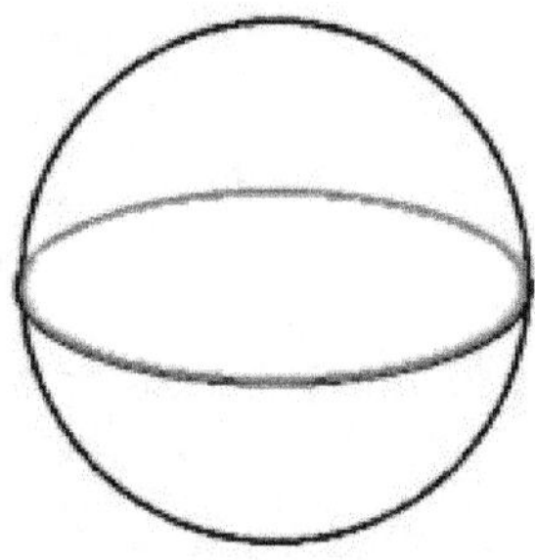

Assume the sphere to have a radius $= r$

Surface area of sphere $= 4\pi r^2$

Volume of sphere $= (4/3)\pi r^3$

The Cylinder

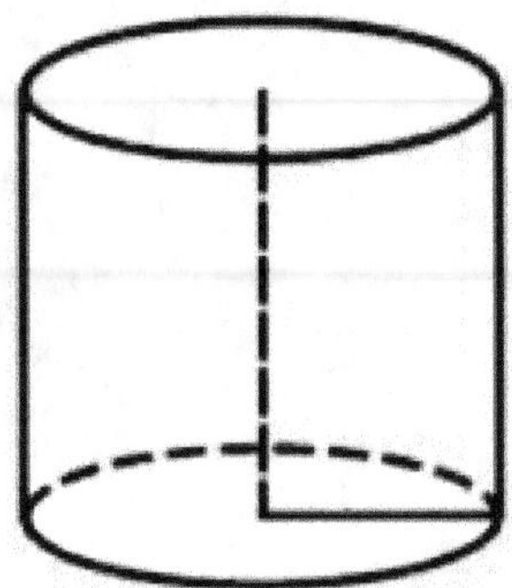

Assume the cylinder to have a radius = r and a height = h

Surface area of cylinder = $2\pi r^2 + 2\pi rh$

Volume of cylinder = $\pi r^2 h$

The Cone

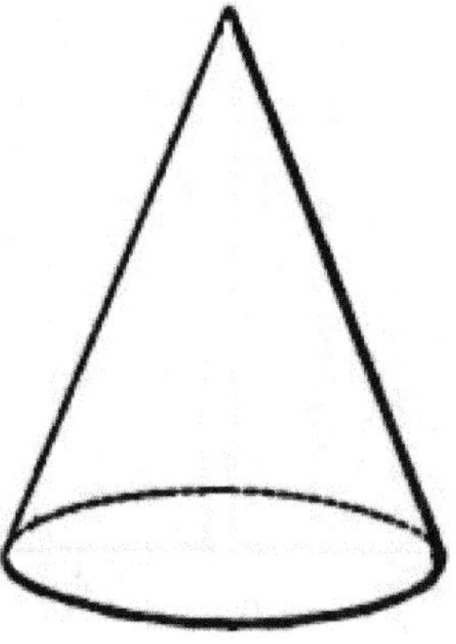

Assume the cone to have a radius = r and a height = h

Surface area of cone = $\pi r^2 + \frac{1}{2}(2\pi rl)$ where l is the inclined height.

Volume of cone = $(1/3)\pi r^2 h$

Note that the inclined height $l = \sqrt{r^2 + h^2}$

The Triangular Prism

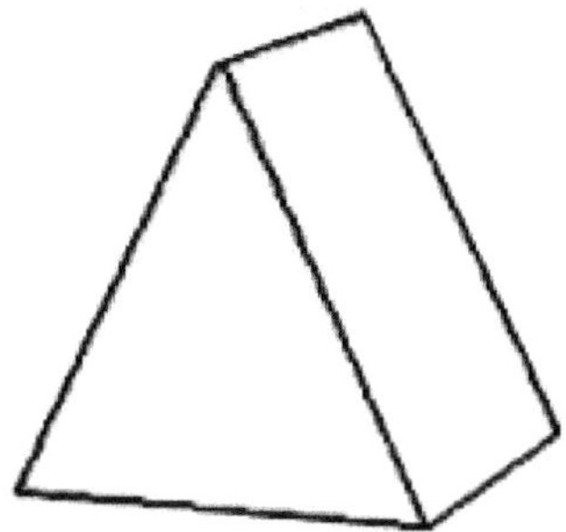

Assume the prism to have a height h and a triangular base of sides a, b, c

Surface area of prism = $2(1/2\ x.a) + (a+b+c)h$ where x is the perpendicular from one vertex of the base to its opposite side.

Volume of prism = $(1/2)\ x.ah$

For a triangular base that is equilateral, that is all sides are equal with length s, then we have a+b+c = 3s.

In this special case, the perpendicular length x =

$$\sqrt{s^2 - \left(\frac{s}{2}\right)^2} = \frac{\sqrt{3}}{2}s$$

The Rectangular Prism

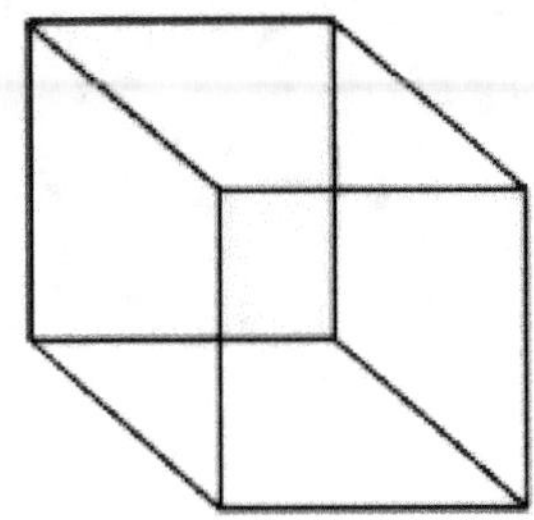

Assume the prism to have sides a,b,c

Surface area of prism = 2ab+2ac+2bc

Volume of prism = abc

The Pentagonal Prism

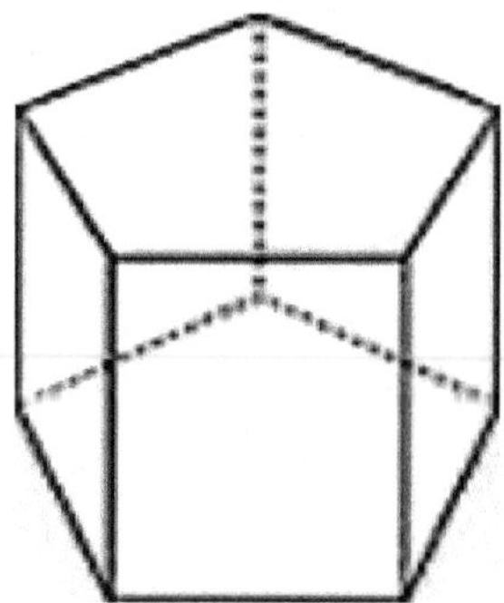

Assume the prism to have a height h and a regular pentagon as a base with side s

Surface area of prism = 5sh + 2(1/2 x.s)(5) = 5sh + 5x.s where x is the perpendicular length from the center of the base to one of the sides.

Volume of prism = ½ (x.s)(5)h = (5/2) s.x.h

where x = $\sqrt{s^2 - \left(\dfrac{s}{2}\right)^2} = \dfrac{\sqrt{3}}{2}s$.

The Hexagonal Prism

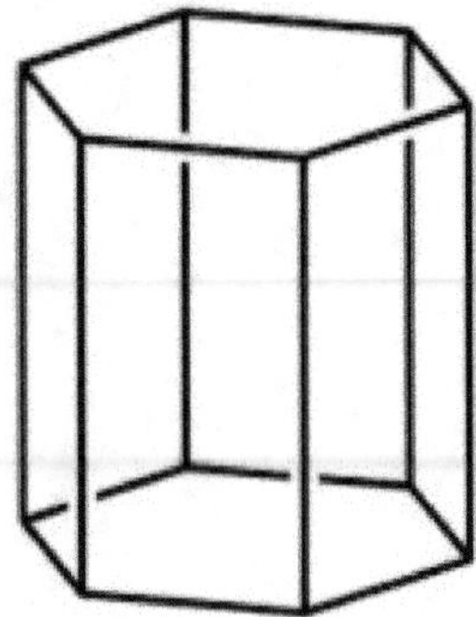

Assume the prism to have a height h and a regular hexagon as a base with side s

Surface area of prism = 6sh + 2(1/2 x.s)(6) = 6sh + 6x.s where x is the perpendicular length from the center of the base to one of the sides.

Volume of prism = ½ (x.s)(6)h = (6/2) s.x.h , where x =

$$\sqrt{s^2 - \left(\frac{s}{2}\right)^2} = \frac{\sqrt{3}}{2}s$$
.

The Triangular Pyramid

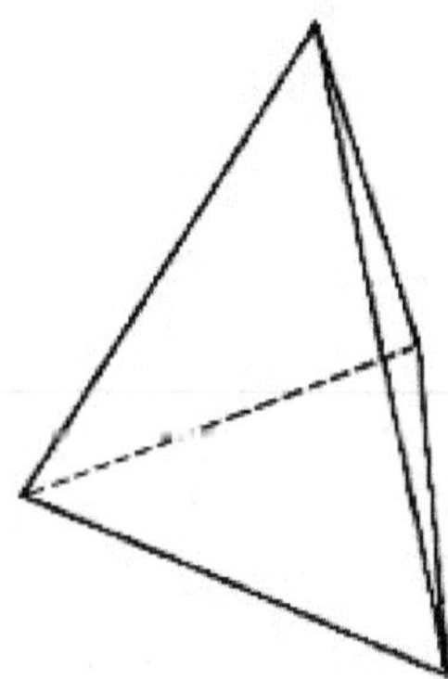

Assume the pyramid to have a height h and a triangular base of equilateral sides with side s

Surface area of pyramid = 3(1/2 s.l) + ½ s.x where x is the perpendicular from one vertex of the base to its opposite side, and l is the inclined height.

Volume of pyramid = (1/3) (1/2 x.s)h = (1/6) x.sh

where the perpendicular length $x = \sqrt{s^2 - \left(\dfrac{s}{2}\right)^2} = \dfrac{\sqrt{3}}{2}s$

and the inclined length $l = \sqrt{x^2 + h^2}$

The Square Pyramid

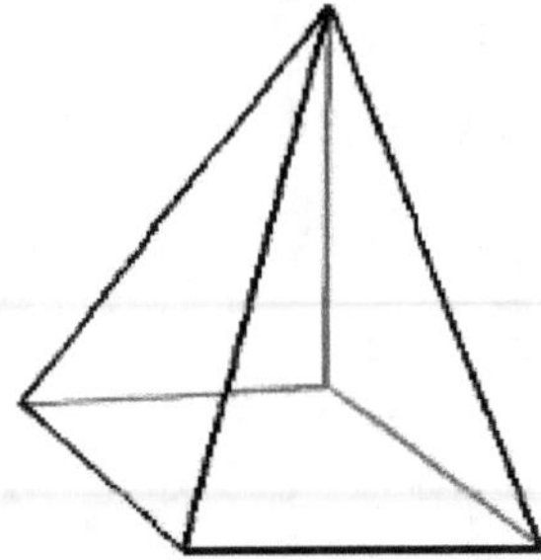

Assume the pyramid to have height h and a square base of side s

Surface area of pyramid $= 4(1/2\ s.l) + s^2$

Volume of pyramid $= (1/3)\ s^2h$

where the inclined length $l = \sqrt{x^2 + h^2}$

The Pentagonal Pyramid

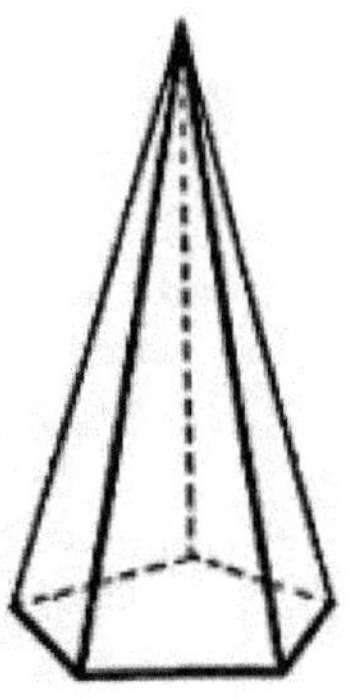

Assume the pyramid to have a height h and a regular pentagon as a base with side s

Surface area of pyramid = ½ (s.l)(5) + ½ (x.s)(5) where x is the perpendicular length from the center of the base to one of the sides, and l is the inclined height.

Volume of pyramid = (1/3)(5)(1/2 x.s)h

where x = $\sqrt{s^2 - \left(\dfrac{s}{2}\right)^2} = \dfrac{\sqrt{3}}{2} s$

and the inclined length l = $\sqrt{x^2 + h^2}$

The Hexagonal Pyramid

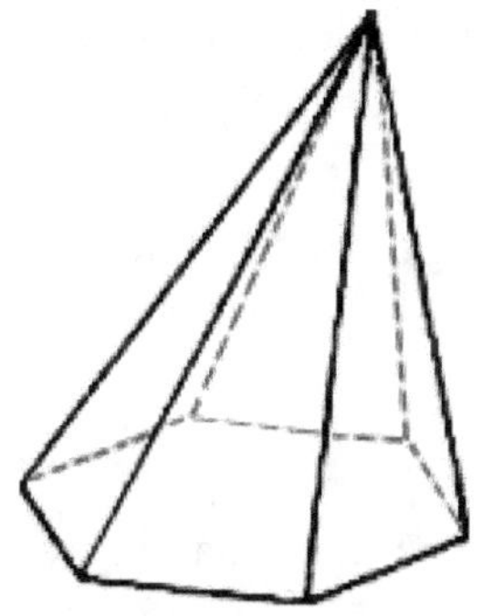

Assume the pyramid to have a height h and a regular hexagon as a base with side s

Surface area of pyramid = ½ (s.l)(6) + ½ (x.s)(6) where x is the perpendicular length from the center of the base to one of the sides, and l is the inclined height.

Volume of pyramid = (1/3)(6)(1/2 x.s)h

where x = $\sqrt{s^2 - \left(\dfrac{s}{2}\right)^2} = \dfrac{\sqrt{3}}{2}s$

and the inclined length l = $\sqrt{x^2 + h^2}$

The Tetrahedron

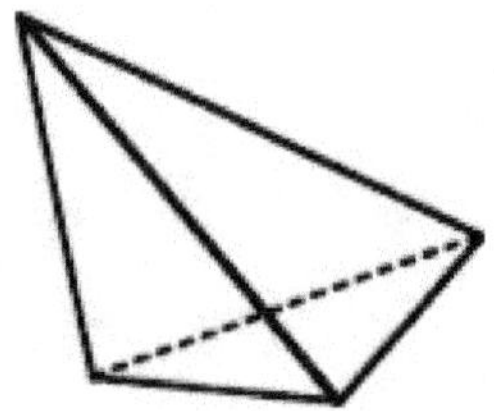

The tetrahedron is a solid composed of four identical faces. Each face is an equilateral triangle of side s.

Surface area of tetrahedron = $\sqrt{3}s^2$

Volume of tetrahedron = $\dfrac{\sqrt{2}}{12}s^3$

Note that the tetrahedron is the same as the triangular pyramid with equal faces.

The Octahedron

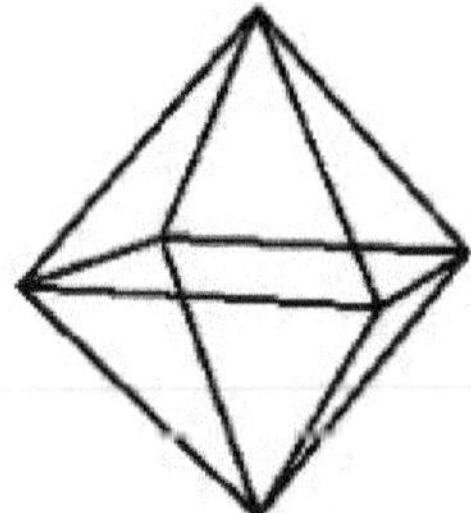

The octahedron is a solid composed of eight identical faces. Each face is an equilateral triangle of side s.

Surface area of octahedron $= 2\sqrt{3}s^2$

Volume of tetrahedron $= \dfrac{\sqrt{2}}{3}s^3$

The Dodecahedron

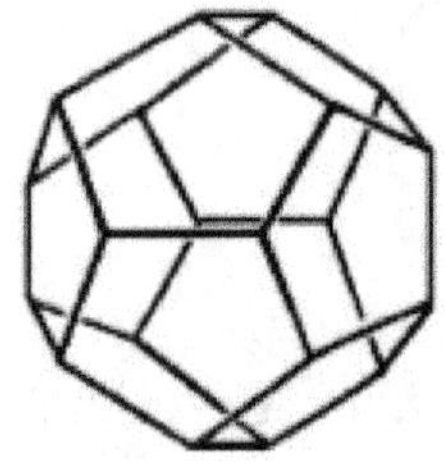

The dodecahedron is a solid composed of twelve identical faces. Each face is a regular pentagon of side s.

Surface area of dodecahedron $= 3\sqrt{25+10\sqrt{5}}s^2$

Volume of dodecahedron = $\dfrac{15+7\sqrt{5}}{4}s^3$

The Icosahedron

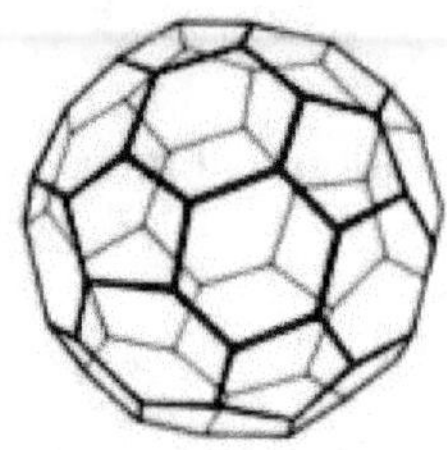

The icosahedron is a solid composed of twenty identical faces. Each face is an equilateral triangle of side s.

Surface area of icosahedron = $5\sqrt{3}s^2$

Volume of icosahedron = $\dfrac{5}{12}\left(3+\sqrt{5}\right)s^3$

References

1. Ratner, M. A. and Ratner, D., *Nanotechnology: A Gentle Introduction to the Next Big Idea*, Prentice Hall, First Edition, 2002.

2. Boysen, E., and Muir, N. C., *Nanotechnology for Dummies*, For Dummies Series, Second Edition, 2011.

3. Binns, C., *Introduction to Nanoscience and Nanotechnology*, Wiley, 2010.

4. Williams, L. and Adams, W., *Nanotechnology Demystified*, McGraw-Hill, 2006.

5. Drexler, K. E., *Engines of Creation: The Coming Era of Nanotechnology*, Anchor, 1987.

6. Foster, L. E., *Nanotechnology: Science, Innovation, and Opportunity*, Prentice Hall, 2005.

7. Wolf, E. L., *Nanophysics and Nanotechnology: An Introduction to Modern Concepts in Nanoscience*, Wiley-VCH, Second Edition, 2006.

8. Mongillo, J. F. *Nanotechnology 101*, Greenwood, 2007.

9. Wilson, M., Kannangara, K., Smith, G., and Simmons, M., *Nanotechnology: Basic Science and Emerging Technologies*, Chapman and Hall/CRC, First Edition, 2002.

10. Horrnyak, G. L. and Dutta, J., Tibbals, H. F., and Rao, A., *Introduction to Nanoscience*, CRC Press, First Edition, 2008.

11. Sanghera, P., Gateway to Nanotechnology: *An Introduction to Nanotechnology for Beginner Students and Professionals*, Booksurge Publishing, 2009.

12. Bhushan, B., *Springer Handbook of Nanotechnology*, Springer, Third Edition, 2010.

13. Lindsay, S., *Introduction to Nanoscience*, Oxford University Press, 2009.

14. Wikipedia – The Free Internet Encyclopedia, www.wikipedia.org

Footnotes

[1] No units are used in these illustrative examples.
[2] No units are used in these illustrative examples.
[3] No units are used in these illustrative examples.
[4] No units are used in these illustrative examples.

2. Chaos Theory Simply Explained

This is a concise article detailing mathematical equations showing when and where chaos occurs in linear algebraic equations and systems of linear simultaneous algebraic equations.

The approach used is very simple and easy to understand by readers. Instead of using quadratic equations like the logistic equation to study chaos theory, we use simple linear equations for this purpose. The article starts with a single simple linear equation and ends with a system of two simultaneous linear equations.

The conditions under which chaos occurs in linear equations are precisely investigated using several examples. The derivation of all necessary equations is shown in great detail. This article is aimed at the novice who is just starting his or her study of chaos and chaos theory. The reader is expected to know the basics of algebra at an elementary level.

What is Chaos?

Most books on chaos theory start with what is called the logistic equation[3]. This is a nonlinear equation (actually quadratic) that exhibits very complicated phenomena when iterated. The logistic equation is mostly used in modeling population growth and other processes. However, this article will take a simple approach. The logistic equation will not be used here to illustrate chaos. What will be used is a much simpler linear equation for this purpose. But before we do that,

---

[3] The logistic equation is $y = ax(1 - x)$ where a is a constant.

let us define precisely what we mean by chaos and chaos theory.

According to the dictionary.com website, chaos is "a state of utter confusion and disorder; a total lack of organization or order." According to the wikipedia.com website (The Free Encyclopedia), chaos theory is "a field of study in mathematics, with applications in several disciplines including physics, economics, biology, and philosophy. Chaos theory studies the behavior of dynamical systems that are highly sensitive to initial conditions, an effect which is popularly referred to as the butterfly effect. Small differences in initial conditions yield widely diverging outcomes for chaotic systems, rendering long-term prediction impossible in general. This happens even though these systems are deterministic, meaning that their future behavior is fully determined by their initial conditions, with no random elements involved."

In this article, we will illustrate the meaning of the above quote from wikipedia.com in full detail. Using a simple linear equation, we will illustrate how and when chaos occurs. We will show also that chaotic systems are usually unpredictable even though their behavior is fully determined by the governing equations and initial conditions. Examples of chaotic systems are the weather, the stock market, and earthquakes.

Let me first give you an idea about what we mean by chaos in the mathematical sense. Suppose we start with a process (mathematical or natural) or an equation with a certain number and end up with a final number. The number we start with is called the initial condition and the number we end with is the result. Suppose we start with 1 and end up with 10. Let us now change the initial condition slightly and start with 1.1 instead of 1. If we go through the same process mathematically, suppose we get a result close to 10, e.g.

suppose we obtain 10.3 or 10.5. In this case everything would be normal and predictable. Obviously, there is no chaos.

However, suppose that when we start with 1.1 we end up with 12 or 15 or even 20. In this case, the result is totally different from 10 and there is a huge difference. Changing the initial condition slightly from 1 to 1.1, results in two totally different answers. Clearly, this is not predictable. This is exactly what we mean by chaos. In this case, we say that the process is chaotic.

Examples of Chaos

Let us start with a simple linear equation in the following form:

$$y = x - 1$$

In the above equation, we have two variables; namely x and y. The variable x is called the independent variable and the variable is called the dependent variable. We substitute a value for x and obtain the corresponding value for y. For example, when we use the value $x = 0$, using the above equation we obtain $y = 0 - 1 = -1$. When we substitute the value $x = 1$, using the above equation we obtain $y = 1 - 1 = 0$.

What we want to do now is to perform a long process of operations using the above equation that is called the process of iteration. We will start with an initial value for x, then obtain a value for y. Then we use this new value of y to substitute in the equation for x in order to get another value for y. We continue in this fashion, i.e. substitute each obtained value in the equation several times. Let us perform this operation of iteration ten times using a specific initial value for x. But before we do this, let us change the notation used in the equation to make it clearer and easier for iterating x. Since we will using a new value for x every time we use the equation, let us denote the current value by $x(n)$ and the obtained value $x(n+1)$. Let the value of the parameter n range from 0, 1, 2, 3, 4, 5, 6, 7, 8, 9, 10.

So the initial value to start the iteration is called $x(0)$. The next values would be called $x(1), x(2), x(3), x(4), x(5), x(6), x(7), x(8), x(9), x(10)$. Let us obtain these ten values (called iterates) for an initial value $x(0) = 0.2673$. We will explain later why we chose an initial value less than 1.

Let us perform the iteration as follows:

$$x(1) = x(0) - 1 = 0.2673 - 1 = -0.7327$$

$$x(2) = x(1) - 1 = -0.7327 - 1 = -1.7327$$

$$x(3) = x(2) - 1 = -1.7327 - 1 = -2.7327$$

$$x(4) = x(3) - 1 = -2.7327 - 1 = -3.7327$$

$$x(5) = x(4) - 1 = -3.7327 - 1 = -4.7327$$

$$x(6) = x(5) - 1 = -4.7327 - 1 = -5.7327$$

$$x(7) = x(6) - 1 = -5.7327 - 1 = -6.7327$$

$$x(8) = x(7) - 1 = -6.7327 - 1 = -7.7327$$

$$x(9) = x(8) - 1 = -7.7327 - 1 = -8.7327$$

$$x(10) = x(9) - 1 = -8.7327 - 1 = -9.7327$$

In order to find out if there is chaos in this equation (or process), we need to repeat the above iterative procedure but using a slightly changed initial value. Let us change the initial value by a very small number, say 0.0001, i.e. let us use the initial value of 0.2674 instead of 0.2673.

We iterate the above linear equation ten times starting with the initial value 0.2674 as follows:

$$x(1) = x(0) - 1 = 0.2674 - 1 = -0.7326$$

$$x(2) = x(1) - 1 = -0.7326 - 1 = -1.7326$$

$$x(3) = x(2) - 1 = -1.7326 - 1 = -2.7326$$

$$x(4) = x(3) - 1 = -2.7326 - 1 = -3.7326$$

$$x(5) = x(4) - 1 = -3.7326 - 1 = -4.7326$$

$$x(6) = x(5) - 1 = -4.7326 - 1 = -5.7326$$

$$x(7) = x(6) - 1 = -5.7326 - 1 = -6.7326$$

$$x(8) = x(7) - 1 = -6.7326 - 1 = -7.7326$$

$$x(9) = x(8) - 1 = -7.7326 - 1 = -8.7326$$

$$x(10) = x(9) - 1 = -8.7326 - 1 = -9.7326$$

Comparing the above iterations of the simple linear equation $y = x - 1$, we notice that when we changed the initial value slightly from 0.2673 to 0.2674 (a change of 0.0001), the final result after ten iterations changed also slightly from -9.7327 to -9.7326 (a change also of 0.0001). This slight change in the final result is perfectly normal and expected. It is also perfectly predictable. Thus, we say that there is no chaos manifested in this example.

Thus, our conclusion is that the simple linear equation $y = x - 1$ when iterated does not exhibit chaotic behavior. What we will show next is that if we use a slightly different simple linear equation, it will exhibit significant chaotic behavior.

Let us now consider the following equation and repeat the above iterative procedure using exactly the same starting values for x; i.e. the same initial conditions.

$$y = 2x - 1$$

Iterating the above equations in the same way like we did before, one would expect to get similar results but this is not the case here even though the new linear equation is very similar to the previous linear equation except for the coefficient 2.

Let us use the same initial value of 0.2673 and perform the iteration ten times as follows:

$$x(1) = 2x(0) - 1 = 2(0.2673) - 1 = -0.4654$$

$$x(2) = 2x(1) - 1 = 2(-0.4654) - 1 = -1.9308$$

$$x(3) = 2x(2) - 1 = 2(-1.9308) - 1 = -4.8616$$

$$x(4) = 2x(3) - 1 = 2(-4.8616) - 1 = -10.7232$$

$$x(5) = 2x(4) - 1 = 2(-10.7232) - 1 = -22.4464$$

$$x(6) = 2x(5) - 1 = 2(-22.4464) - 1 = -45.8928$$

$$x(7) = 2x(6) - 1 = 2(-45.8928) - 1 = -92.7856$$

$$x(8) = 2x(7) - 1 = 2(-92.7856) - 1 = -186.5712$$

$$x(9) = 2x(8) - 1 = 2(-186.5712) - 1 = -374.1424$$

$$x(10) = 2x(9) - 1 = 2(-374.1424) - 1 = -749.2848$$

In order to find out if there is chaos in this equation, we need to repeat the above iterative procedure but using a slightly changed initial value. Let us change the initial value by a very small number, say 0.0001; i.e. let us use the initial value of 0.2674 instead of 0.2673 (as we did before for the previous equation). Here are the ten iterates:

$$x(1) = 2x(0) - 1 = 2(0.2674) - 1 = -0.4652$$

$$x(2) = 2x(1) - 1 = 2(-0.4652) - 1 = -1.9304$$

$$x(3) = 2x(2) - 1 = 2(-1.9304) - 1 = -4.8608$$

$$x(4) = 2x(3) - 1 = 2(-4.8608) - 1 = -10.7216$$

$$x(5) = 2x(4) - 1 = 2(-10.7216) - 1 = -22.4432$$

$$x(6) = 2x(5) - 1 = 2(-22.4432) - 1 = -45.8864$$

$$x(7) = 2x(6) - 1 = 2(-45.8864) - 1 = -92.7728$$

$$x(8) = 2x(7) - 1 = 2(-92.7728) - 1 = -186.5456$$

$$x(9) = 2x(8) - 1 = 2(-186.5456) - 1 = -374.0912$$

$$x(10) = 2x(9) - 1 = 2(-374.0912) - 1 = -749.1824$$

Comparing the above iterations of the simple linear equation $y = 2x - 1$, we notice that when we changed the initial value slightly from 0.2673 to 0.2674 (a change of 0.0001), the final result after ten iterations changed from -749.2848 to -749.1824 (a significant change of 0.1024). This substantial change in the final result is not expected. It is also totally unpredictable. Thus, we say that there is chaos manifested in this example. Thus, the equation $y = 2x - 1$ exhibits chaotic behavior when iterated. Indeed, we will notice that we would obtain totally different results if we perform this simple iteration 100 times or 1000 times. This is the essence of chaos.

The question that we need to answer now is that why the equation $y = 2x - 1$ exhibits chaotic behavior when iterated while the equation $y = x - 1$ does not. Let us examine the characteristics of each equation separately to see why this difference in behavior occurs.

Let us first iterate the equation $y = x - 1$ without using numbers as follows:

$$x(1) = x(0) - 1$$

$$x(2) = x(1) - 1 = (x(0) - 1) - 1 = x(0) - 2$$

$$x(3) = x(2) - 1 = (x(0) - 2) - 1 = x(0) - 3$$

$$x(4) = x(3) - 1 = (x(0) - 3) - 1 = x(0) - 4$$

We continue to iterate the above equation until we can write the general equation for iteration as follows in terms of n where $n = 1, 2, 3, \ldots\ldots$ (this equation is derived in this form based on the above iterations)

$$x(n) = x(0) - n$$

The above equation is an explicit equation for the n-th iterate of x. Let us now consider what happens when we change the initial condition $x(0)$ slightly. Consider the new initial condition $x(0) + \varepsilon$ where ε is a very small number (it was 0.0001 in our numerical example above). Let us call the new iterate in this case $x(m)$ (this is the m-th iterate). Based on the above equation, we get the following expression for the m-th iterate when the new initial condition is used:

$$x(m) = (x(0) + \varepsilon) - m = x(0) + \varepsilon - m$$

Subtracting the above two equations from each other, we obtain:

$$x(m) - x(n) = (x(0) + \varepsilon - m) - (x(0) - n) = \varepsilon + n - m$$

To see if this linear equation is chaotic, then we let $n = m$, to obtain:

$$x(n) - x(n) = \varepsilon + n - n = \varepsilon$$

and since ε is a very small number, the above equation goes to zero as the value of n increases. Thus, in this case, we get equal results for $x(n)$ and $x(m)$ as both n and m increase to infinity.

Let us now repeat the above derivation for the equation $y = 2x - 1$ to see what happens when we iterate it without using numbers. In this case, we obtain the following iterations:

$$x(1) = 2x(0) - 1$$

$$x(2) = 2x(1) - 1 = 2(2x(0) - 1) - 1 = 2^2 x(0) - 2 - 1$$

$$x(3) = 2x(2) - 1 = 2(2^2 x(0) - 2 - 1) - 1 = 2^3 x(0) - 2^2 - 2 - 1$$

$$x(4) = 2x(3) - 1 = 2(2^3 x(0) - 2^2 - 2 - 1) - 1 = 2^4 x(0) - 2^3 - 2^2 - 2 - 1$$

We continue to iterate the above equation until we can write the general equation for iteration as follows in terms of n where $n = 1, 2, 3, \ldots\ldots$ (this equation is derived in this form based on the above iterations)

$$x(n) = 2^n x(0) - 2^{n-1} - 2^{n-2} - 2^{n-3} - \ldots\ldots - 1$$

The above equation can be re-written in the following form:

$$x(n) = 2^n x(0) - \left(1 + 2 + 2^2 + \ldots\ldots + 2^{n-3} + 2^{n-2} + 2^{n-1}\right)$$

We recognize that the expression between the parentheses in the above equation is a geometric series. Reviewing our knowledge from the theory of series and sequences (see the Appendix for details), we can simplify the expression between the parenthesis as follows:

$$1 + 2 + 2^2 + \ldots\ldots + 2^{n-3} + 2^{n-2} + 2^{n-1} = \frac{2^n - 1}{2 - 1} = 2^n - 1$$

Thus, the equation for $x(n)$ can now be written as follows:

$$x(n) = 2^n x(0) - \left(2^n - 1\right)$$

The above equation is an explicit equation for the n-th iterate of x. Let us now consider what happens when we change the initial condition $x(0)$ slightly. Consider the new initial condition $x(0) + \varepsilon$ where ε is a very small number (it was 0.0001 in our numerical example above). Let us call the new iterate in this case $x(m)$ (this is the m-th iterate). Based on the above equation, we get the following expression for the m-th iterate when the new initial condition is used:

$$x(m) = 2^m \left(x(0) + \varepsilon\right) - \left(2^m - 1\right)$$

Subtracting the above two equations from each other, we obtain:

$$x(m) - x(n) = \left[2^m \left(x(0) + \varepsilon\right) - \left(2^m - 1\right)\right] - \left[2^n x(0) - \left(2^n - 1\right)\right]$$

Thus, we obtain after simplifying the resulting equation:

$$x(m) - x(n) = \left(2^m - 2^n\right)x(0) + 2^m \varepsilon + 2^n - 2^m$$

To see if this linear equation is chaotic, then we let $n = m$, to obtain:

$$x(n) - x(n) = 2^n \varepsilon$$

We notice that ε is a very small number, but the above equation does not go to zero as the value of n increases. This is because of the coefficient 2^n that appears in the final result. Actually, this expression goes to infinity as n increases to infinity. Thus, in this case, we do not get equal results for $x(n)$ and $x(m)$ as both n and m increase to infinity. Thus, this equation exhibits chaos.

Comparing the two linear equations, we notice that when the coefficient of x is larger than 1, then chaos occurs. However, when the coefficient of x is equal to 1, then there is no chaos. This is because $1^\varepsilon = 1$ irrespective of the value of the parameter ε. Obviously, this is not the case when the coefficient is 2 or larger. Actually chaos occurs when the coefficient is larger than 1. The question now is: What happens to the iterations when the coefficient is smaller than 1?

Let us consider the linear equation $y = \dfrac{1}{2}x - 1$ where the coefficient of x is less than 1. In this case, and by comparison with the previous derivation, the n-th iterate approaches the quantity $\left(\dfrac{1}{2}\right)^n \varepsilon$, where ε is a very small number. Clearly, this quantity approaches zero as the value of n increases to infinity. Thus, for this linear equation, there is no chaos. In general, the linear equation is not chaotic when the coefficient of x is less than 1.

Next, we will investigate what happens when we iterate several linear equations simultaneously. For simplicity, we will consider the following systems of two linear equations:

$$x_{n+1} = 2x_n + y_n + 1$$
$$y_{n+1} = x_n + 2y_n + 2$$

The above system can be re-written in the following matrix form:

$$\begin{Bmatrix} x_{n+1} \\ y_{n+1} \end{Bmatrix} = \begin{bmatrix} 2 & 1 \\ 1 & 2 \end{bmatrix} \begin{Bmatrix} x_n \\ y_n \end{Bmatrix} + \begin{Bmatrix} 1 \\ 2 \end{Bmatrix}$$

Let us iterate the above system of equations several times to see if chaos occurs in this example. Let us start with the initial condition and make the first iteration as follows:

$$\begin{Bmatrix} x_1 \\ y_1 \end{Bmatrix} = \begin{bmatrix} 2 & 1 \\ 1 & 2 \end{bmatrix} \begin{Bmatrix} x_0 \\ y_0 \end{Bmatrix} + \begin{Bmatrix} 1 \\ 2 \end{Bmatrix}$$

The second iteration is then obtained as follows:

$$\begin{Bmatrix} x_2 \\ y_2 \end{Bmatrix} = \begin{bmatrix} 2 & 1 \\ 1 & 2 \end{bmatrix} \begin{Bmatrix} x_1 \\ y_1 \end{Bmatrix} + \begin{Bmatrix} 1 \\ 2 \end{Bmatrix}$$

Substituting the values of the first iteration in the above equation, we obtain:

$$\left\{\begin{matrix} x_2 \\ y_2 \end{matrix}\right\} = \begin{bmatrix} 2 & 1 \\ 1 & 2 \end{bmatrix}\left(\begin{bmatrix} 2 & 1 \\ 1 & 2 \end{bmatrix}\left\{\begin{matrix} x_0 \\ y_0 \end{matrix}\right\} + \left\{\begin{matrix} 1 \\ 2 \end{matrix}\right\}\right) + \left\{\begin{matrix} 1 \\ 2 \end{matrix}\right\}$$

Simplifying the above equation, we obtain:

$$\left\{\begin{matrix} x_2 \\ y_2 \end{matrix}\right\} = \begin{bmatrix} 2 & 1 \\ 1 & 2 \end{bmatrix}^2\left\{\begin{matrix} x_0 \\ y_0 \end{matrix}\right\} + \begin{bmatrix} 2 & 1 \\ 1 & 2 \end{bmatrix}\left\{\begin{matrix} 1 \\ 2 \end{matrix}\right\} + \left\{\begin{matrix} 1 \\ 2 \end{matrix}\right\}$$

Continuing with the third iteration upon the above equation and simplifying the results, we obtain:

$$\left\{\begin{matrix} x_3 \\ y_3 \end{matrix}\right\} = \begin{bmatrix} 2 & 1 \\ 1 & 2 \end{bmatrix}^3\left\{\begin{matrix} x_0 \\ y_0 \end{matrix}\right\} + \begin{bmatrix} 2 & 1 \\ 1 & 2 \end{bmatrix}^2\left\{\begin{matrix} 1 \\ 2 \end{matrix}\right\} + \begin{bmatrix} 2 & 1 \\ 1 & 2 \end{bmatrix}\left\{\begin{matrix} 1 \\ 2 \end{matrix}\right\} + \left\{\begin{matrix} 1 \\ 2 \end{matrix}\right\}$$

In general, we obtain the following formula after performing the nth iteration:

$$\left\{\begin{matrix} x_n \\ y_n \end{matrix}\right\} = \begin{bmatrix} 2 & 1 \\ 1 & 2 \end{bmatrix}^n\left\{\begin{matrix} x_0 \\ y_0 \end{matrix}\right\} + \begin{bmatrix} 2 & 1 \\ 1 & 2 \end{bmatrix}^{n-1}\left\{\begin{matrix} 1 \\ 2 \end{matrix}\right\} + \begin{bmatrix} 2 & 1 \\ 1 & 2 \end{bmatrix}^{n-2}\left\{\begin{matrix} 1 \\ 2 \end{matrix}\right\} +$$

$$+ \cdots\cdots + \begin{bmatrix} 2 & 1 \\ 1 & 2 \end{bmatrix}\left\{\begin{matrix} 1 \\ 2 \end{matrix}\right\} + \left\{\begin{matrix} 1 \\ 2 \end{matrix}\right\}$$

Next, we will re-write the above equation in the following more simplified form:

$$\left\{\begin{matrix} x_n \\ y_n \end{matrix}\right\} = \begin{bmatrix} 2 & 1 \\ 1 & 2 \end{bmatrix}^n\left\{\begin{matrix} x_0 \\ y_0 \end{matrix}\right\} + \left(\begin{bmatrix} 2 & 1 \\ 1 & 2 \end{bmatrix}^{n-1} + \begin{bmatrix} 2 & 1 \\ 1 & 2 \end{bmatrix}^{n-2} + \cdots\cdots + \begin{bmatrix} 2 & 1 \\ 1 & 2 \end{bmatrix} + \begin{bmatrix} 1 & 0 \\ 0 & 1 \end{bmatrix}\right)\left\{\begin{matrix} 1 \\ 2 \end{matrix}\right\}$$

The expression inside the parenthesis is a geometric series of the following form:

$$[A]^{n-1} + [A]^{n-2} + \ldots\ldots + [I]$$

In order to simplify the above expression, we need to use the following mathematical identity:

$$\left([A]^{n-1} + [A]^{n-2} + \ldots\ldots + [I]\right)\left([I] - [A]\right) = [I] - [A]^{n}$$

Therefore, the nth iteration of the linear system of equations becomes:

$$\begin{Bmatrix} x_n \\ y_n \end{Bmatrix} = \begin{bmatrix} 2 & 1 \\ 1 & 2 \end{bmatrix}^{n} \begin{Bmatrix} x_0 \\ y_0 \end{Bmatrix} + \left(\begin{bmatrix} 1 & 0 \\ 0 & 1 \end{bmatrix} - \begin{bmatrix} 2 & 1 \\ 1 & 2 \end{bmatrix}^{n}\right)\left(\begin{bmatrix} 1 & 0 \\ 0 & 1 \end{bmatrix} - \begin{bmatrix} 2 & 1 \\ 1 & 2 \end{bmatrix}\right)^{-1} \begin{Bmatrix} 1 \\ 2 \end{Bmatrix}$$

The above equation is an explicit equation for the n-th iterate of $\begin{Bmatrix} x \\ y \end{Bmatrix}$. Let us now consider what happens when we change the initial condition $\begin{Bmatrix} x(0) \\ y(0) \end{Bmatrix}$ slightly. Consider the new initial condition $\begin{Bmatrix} x(0) + \varepsilon \\ y(0) + \delta \end{Bmatrix}$ where ε and δ are very small numbers. Let us call the new iterate in this case $\begin{Bmatrix} x_m \\ y_m \end{Bmatrix}$ (this is the m-th iterate). Based on the above equation, we get the following expression for the m-th iterate when the new initial condition is used:

$$\begin{Bmatrix} x_m \\ y_m \end{Bmatrix} = \begin{bmatrix} 2 & 1 \\ 1 & 2 \end{bmatrix}^{m} \begin{Bmatrix} x_0 + \varepsilon \\ y_0 + \delta \end{Bmatrix} + \left(\begin{bmatrix} 1 & 0 \\ 0 & 1 \end{bmatrix} - \begin{bmatrix} 2 & 1 \\ 1 & 2 \end{bmatrix}^{m}\right)\left(\begin{bmatrix} 1 & 0 \\ 0 & 1 \end{bmatrix} - \begin{bmatrix} 2 & 1 \\ 1 & 2 \end{bmatrix}\right)^{-1} \begin{Bmatrix} 1 \\ 2 \end{Bmatrix}$$

Subtracting the above two equations from each other, we obtain:

$$\begin{Bmatrix} x_m \\ y_m \end{Bmatrix} - \begin{Bmatrix} x_n \\ y_n \end{Bmatrix} = \begin{bmatrix} 2 & 1 \\ 1 & 2 \end{bmatrix}^m \begin{Bmatrix} x_0 + \varepsilon \\ y_0 + \delta \end{Bmatrix} - \begin{bmatrix} 2 & 1 \\ 1 & 2 \end{bmatrix}^n \begin{Bmatrix} x_0 \\ y_0 \end{Bmatrix} + \left\langle \begin{matrix} \left(\begin{bmatrix} 1 & 0 \\ 0 & 1 \end{bmatrix} - \begin{bmatrix} 2 & 1 \\ 1 & 2 \end{bmatrix}^m \right) \left(\begin{bmatrix} 1 & 0 \\ 0 & 1 \end{bmatrix} - \begin{bmatrix} 2 & 1 \\ 1 & 2 \end{bmatrix} \right)^{-1} - \\ \left(\begin{bmatrix} 1 & 0 \\ 0 & 1 \end{bmatrix} - \begin{bmatrix} 2 & 1 \\ 1 & 2 \end{bmatrix}^n \right) \left(\begin{bmatrix} 1 & 0 \\ 0 & 1 \end{bmatrix} - \begin{bmatrix} 2 & 1 \\ 1 & 2 \end{bmatrix} \right)^{-1} \end{matrix} \right\rangle \begin{Bmatrix} 1 \\ 2 \end{Bmatrix}$$

To see if this linear equation is chaotic, then we let $n = m$, to obtain:

$$\begin{Bmatrix} x_n \\ y_n \end{Bmatrix} - \begin{Bmatrix} x_n \\ y_n \end{Bmatrix} = \begin{bmatrix} 2 & 1 \\ 1 & 2 \end{bmatrix}^n \begin{Bmatrix} \varepsilon \\ \delta \end{Bmatrix}$$

We notice that ε and δ are very small numbers, but the above equation does not go to zero as the value of n increases. This is because of the coefficient matrix $\begin{bmatrix} 2 & 1 \\ 1 & 2 \end{bmatrix}^n$ that appears in the final result. Actually, this expression goes to infinity as n increases to infinity. The reason is because of the eigenvalues[4] of the matrix $\begin{bmatrix} 2 & 1 \\ 1 & 2 \end{bmatrix}$. If one or more of the eigenvalues of the matrix is greater than 1, then this expression goes to infinity. On the other hand, if the eigenvalues of the matrix are less than one, then the expression goes to zero. For the specific matrix

---

[4] The eigenvalues of a matrix are its characteristic values. They can be obtained by solving the corresponding equation of the characteristic polynomial.

$$\begin{bmatrix} 2 & 1 \\ 1 & 2 \end{bmatrix}$$ in this example, one of the eigenvalues is 3 which is greater than 1 (Check the Appendix for the computation of this value). Thus, in this case, we do not get equal results for $x(n)$ and $x(m)$ as both n and m increase to infinity. Thus, this equation exhibits chaos.

In general, for a system of linear simultaneous algebraic equations, chaos occurs or not depending on the coefficient matrix. If the value of one or more of the eigenvalues is greater than 1, then the system is chaotic. On the other hand, if all the eigenvalues of the coefficient matrix have values less than 1, then the system is not chaotic. For complete details and the mathematical derivation and proof, check the book by Scheinerman entitled "Invitation to Dynamic Systems." (reference [1]).

References

1. Scheinrman, E. R., Invitation to Dynamical Systems, Prentice Hall, 1996.

2. Gleick, J., Chaos: Making a New Science, Revised Edition, Penguin, 2008.

3. Strogatz, S. H., Nonlinear Dynamics and Chaos, With Applications to Physics, Biology, Chemistry, and Engineering, Westview Press, 2001.

4. Smith, L. and Smith, L., Chaos: A Very Short Introduction, Oxford University Press, 2007.

5. Murphy, R. P., Chaos Theory, Ludwig von Mises Institute, 2010.

6. Peitgen, H.-O., Chaos and Fractals: New Frontiers of Science, Second Edition, Springer, 2004.

7. Williams, G. P., Chaos Theory Tamed, CRC Press, 1997.

8. Hilborn, R. C. Chaos and Nonlinear Dynamics: An Introduction for Scientists and Engineers, Second Edition, Oxford University Press, 2001.

9. Schroeder, M., Fractals, Chaos, Power Laws: Minutes from an Infinite Paradise, Dove Publications, 2009.

10. Devaney, R., A First Course in Chaotic Dynamical Systems, Westview Press, 1992.

11. Gribbin, J., Deep Simplicity, Rando House, 2005.

Mathematical Equations Used

Geometric Series

$$1 + a + a^2 + \ldots\ldots + a^{n-3} + a^{n-2} + a^{n-1} = \frac{a^n - 1}{a - 1}, \quad a \neq 0$$

In terms of matrices, we obtain:

$$\left([A]^{n-1} + [A]^{n-2} + \ldots\ldots + [I]\right)\left([I] - [A]\right) = [I] - [A]^n$$

Eignevalues of a Matrix

Calculation of the eignevalues of the matrix $\begin{bmatrix} 2 & 1 \\ 1 & 2 \end{bmatrix}$

Solve the equation $\det(A - \lambda I) = 0$

$$\det\left(\begin{bmatrix} 2 & 1 \\ 1 & 2 \end{bmatrix} - \lambda \begin{bmatrix} 1 & 0 \\ 0 & 1 \end{bmatrix}\right) = 0$$

$$\det\begin{bmatrix} 2 - \lambda & 1 \\ 1 & 2 - \lambda \end{bmatrix} = 0$$

$$(2 - \lambda)^2 - 1 = 0$$

$$4 - 4\lambda + \lambda^2 - 1 = 0$$

$$\lambda^2 - 4\lambda + 3 = 0$$

$$\lambda = 1 \ \ or \ \ 3$$

3. Solving Equations

In this chapter[5] we discuss how to solve algebraic equations using MATLAB. We will discuss both linear and nonlinear algebraic equations. We will also discuss systems of simultaneous linear and nonlinear algebraic equations. First, let us solve the following simple linear equation for the variable x:

$$2x - 3 = 0$$

The solution of the above equation may be obvious to the reader but we will show how to solve it using MATLAB. The easiest method to do it in MATLAB would be to consider the left-hand-side of the equation as a polynomial of first degree and then find the roots of the polynomial using the MATLAB command `roots`. First, we write the above equation as a polynomial as follows:

$$p(x) = 2x - 3 = 0$$

We enter the coefficients of the above polynomial in a vector in MATLAB then use the `roots` command to find the root which is the solution to the above linear equation in x. Here are the needed commands:

```
>> p = [2 -3]

p =

     2    -3

>> roots(p)
```

---

[5] This article appeared originally as a chapter in the book "MATLAB for Beginners – A Gentle Approach."

```
ans =

    1.5000
```

It is clear that the correct solution for x is obtained which is 1.5. Next, let us show similarly how to find the solution of a quadratic equation in x by solving the following quadratic equation:

$$5x^2 + 3x - 4 = 0$$

We will consider the above equation as a polynomial of second degree in x. We will enter the coefficients of the polynomial in a vector in MATLAB followed by the `roots` command. Here are the needed MATLAB commands:

```
>> p = [5 3 -4]

p =

     5     3    -4

>> roots(p)

ans =

   -1.2434
    0.6434
```

It is clear from the above that two real solutions are obtained for x. Let us now solve a more complicated equation. Let us solve the following equation for x:

$$2x^5 - 3x^4 - 5x^3 + x^2 - 1 = 0$$

First, we consider the left-hand-side as a polynomial of fifth degree in x. Then we write the coefficients of this polynomial as a vector in MATLAB followed by the `roots` command as follows (note that we will enter 0 for the missing x in the above polynomial):

```
>> p = [2 -3 -5 1 0 -1]

p =

        2    -3    -5     1     0    -1

>> roots(p)

ans =

   2.4507
  -1.0000
  -0.6391
   0.3442 + 0.4480i
   0.3442 - 0.4480i
```

Note from the above that five values are obtained as the solution of the above equation. These values are the five roots of the polynomial given above. Note that three of these roots are real while two roots are complex conjugates of each other.

Next, let us discuss how to solve a system of linear simultaneous algebraic equations. In order to solve such systems there are several methods available with most of them involving the use of matrices and vectors. Consider the following simple system of two linear algebraic equations in the variables x and y:

$$x - 3y = 5$$
$$4x + 6y = 3$$

In order to solve the above system of equations, we need to re-write the system as a matrix equation. The coefficients of x and y on the left-hand-side of the equations are considered as the elements of a square matrix (of size 2x2 here) while the numbers on the right-hand-side are entered into a vector. The resulting equivalent matrix equation is written as follows:

$$\begin{bmatrix} 1 & -3 \\ 4 & 6 \end{bmatrix} \begin{Bmatrix} x \\ y \end{Bmatrix} = \begin{Bmatrix} 5 \\ 3 \end{Bmatrix}$$

The above matrix equation is in the form $[A]\{x\} = \{b\}$. The solution[6] of this equation is $\{x\} = [A]^{-1}\{b\}$. Thus we need to find the inverse of the coefficient matrix $[A]$ and multiply it with the constant vector $\{b\}$. Here are the needed MATLAB commands along with the solution:

```
>> A = [1 -3 ; 4 6]

A =

     1      -3
     4       6

>> b = [5 ; 3]

b =

     5
```

---

[6] Consult a book on linear algebra for the proof of this solution.

```
       3

>> x = inv(A)*b

x =

   2.1667
  -0.9444
```

Thus it is clear from the above that $x = 2.1667$ and $y = -0.9444$ are the solutions to the above system of equations. Note that in order to solve the above system we had to find the inverse of a 2x2 matrix. In this case, it was quick to find this inverse because the coefficient matrix was small. But for larger matrices, finding the inverse using MATLAB may take more time. Therefore, it is advised to use another method to solve the above system. One such method that is rather fast in execution is called Gaussian elimination. This method is already implemented in MATLAB as matrix division using the backslash operator "\". Here is the solution of the above system again using Gaussian elimination and the backslash operator:

```
>> A = [1 -3 ; 4 6]

A =

     1     -3
     4      6

>> b = [5 ; 3]

b =

     5
     3
```

```
>> x = A\b

x =

    2.1667
   -0.9444
```

It is clear that we obtain exactly the same solution as before but with more speed on the part of MATLAB. The use of Gaussian elimination along with the backslash operator in MATLAB is greatly recommended for solving large systems of algebraic linear simultaneous equations. As an example, let us solve the following system of five algebraic linear simultaneous equations:

$$2x_1 - 4x_2 - x_3 + 3x_4 - x_5 = 3$$
$$x_1 + x_2 - 2x_3 + x_5 = 6$$
$$-x_1 - 3x_2 + x_4 + 3x_5 = -4$$
$$3x_1 - x_2 - x_3 + 4x_4 - x_5 = 1$$
$$x_1 + x_2 - x_3 + 2x_4 = 5$$

First, we re-write the above system as a matrix equation as follows:

$$\begin{bmatrix} 2 & -4 & -1 & 3 & -1 \\ 1 & 1 & -2 & 0 & 1 \\ -1 & -3 & 0 & 1 & 3 \\ 3 & -1 & -1 & 4 & -1 \\ 1 & 1 & -1 & 2 & 0 \end{bmatrix} \begin{Bmatrix} x_1 \\ x_2 \\ x_3 \\ x_4 \\ x_5 \end{Bmatrix} = \begin{Bmatrix} 3 \\ 6 \\ -4 \\ 1 \\ 5 \end{Bmatrix}$$

The following are the MATLAB commands using Gaussian elimination along with the solution:

```
>> A = [2 -4 -1 3 -1 ; 1 1 -2 0 1 ; -1
-3 0 1 3 ; 3 -1 -1 4 -1 ; 1 1 -1 2 0]

A =

     2    -4    -1     3    -1
     1     1    -2     0     1
    -1    -3     0     1     3
     3    -1    -1     4    -1
     1     1    -1     2     0

>> b = [3 ; 6 ; -4 ; 1 ; 5]

b =

     3
     6
    -4
     1
     5

>> x = A\b

x =

   -4.3571
    0.3571
   -6.4286
    1.2857
   -2.8571
```

It is seen from the above result that five real solutions are obtained for the above system of linear equations.

Systems of simultaneous nonlinear algebraic equations can also be solved in MATLAB but these are significantly more difficult to solve. The reason is that there are no direct commands to solve these complicated systems in MATLAB. One will have to use the MATLAB Optimization Toolbox to find certain functions for the solution of these types of equations – for example the command `fsolve` in this toolbox will solve a system of nonlinear equations. Another option is to use the MATLAB Symbolic Math Toolbox to solve such nonlinear systems - using the MATLAB command `solve`. This option will be illustrated in detail below.

Solving Equations with the MATLAB Symbolic Math Toolbox

The MATLAB Symbolic Math Toolbox can be used to solve algebraic equations in MATLAB. In addition, systems of linear and nonlinear algebraic equations can also be solved using this toolbox. There are special commands in this toolbox for solving equations – in particular the MATLAB command `solve`. Please note that the command `roots` that was discussed earlier in this chapter cannot be used for the symbolic solution of algebraic equations. It needs to be replaced with the `solve` command. The use of the MATLAB command `solve` will be illustrated in this section with several examples.

Consider the following linear algebraic symbolic equation that we need to solve for the variable x in terms of the constants a and b:

$$ax + b = 0$$

We will use the MATLAB command `solve` to solve the above equation. Note that we cannot use the trick of writing it as a polynomial and use the `roots`[7] command. Here

---

[7] The `roots` command can only be used with numerical computations.

is the needed format to use the `solve` command in order to solve the above equation:

```
>> syms x
>> syms a
>> syms b
>> x = solve('a*x+b = 0')

x =

-b/a
```

It is noted that the correct solution is obtained which is $x = -\dfrac{b}{a}$. Next, let us solve the following quadratic equation for x in terms of the constants a, b, and c:

$$ax^2 + bx + c = 0$$

The following is the correct format of the `solve` command in order to obtain the two solutions of the above nonlinear equation:

```
>> syms x
>> syms a
>> syms b
>> syms c
>> x = solve('a*x^2 + b*x + c = 0')

x =

 1/2/a*(-b+(b^2-4*a*c)^(1/2))
 1/2/a*(-b-(b^2-4*a*c)^(1/2))
```

It is clear that we obtain the correct two solutions[8] of the quadratic formula. Let us now solve another nonlinear equation not involving a polynomial. Consider the following equation:

$$e^x = \sin x$$

The above equation is a highly nonlinear equation in x. Below is the needed format of the `solve` command to find this solution:

```
>> syms x
>> x = solve('exp(x)-sin(x) = 0')

x =

.36270205612105111082838863639609-
1.1337459194137525371167081219057*i
```

It is clear from the above that we obtained one complex solution of the above equation. The resulting solution is approximated as `0.363 - 1.13i`. At this step, one may use the MATLAB command `format short` to display the solution using four decimal digits only.

Next, we will consider systems of equations. Consider the following system of linear simultaneous algebraic equations:

$$ax - by = c$$
$$2ax + b = 9c$$

---

[8] We can use the MATLAB command `pretty` to display the solution in a nice way but this will not be done here. The correct use of the `pretty` command in the example above would be `pretty(x)`.

The above system of linear equations can also be solved for the variables x and y in terms of the constants a, b, and c using the MATLAB `solve` command as follows:

```
>> syms x
>> syms y
>> syms a
>> syms b
>> syms c
>> [x,y] = solve('a*x - b*y - c = 0', '2*a*x
+ b - 9*c = 0')

x =

-1/2/a*(b-9*c)



y =

-1/2*(-7*c+b)/b
```

It is clear that two solutions are obtained for x and y in terms of a, b, and c as shown above. Our final example will illustrate the solution of a nonlinear system of simultaneous algebraic equations. Consider the following system:

$$x^3 - xy = -3$$
$$x^2 - 2y^2 = 5$$

The above system is a highly nonlinear system. Its solution using the MATLAB command `solve` is illustrated below:

```
>> syms x
>> syms y
>> [x,y] = solve('x^3 - x*y + 3 = 0', 'x^2 -2*y^2
- 5 = 0')
```

```
x =

1.1146270688593392975541180081115+1.36934983263
00382040739329085074*i


.36979370624989001773301777958373+1.06182311951
36687011590573146631*i
  -
1.4844207751092293152871357876952+.282946327302
94451032613534596558*i
 -1.4844207751092293152871357876952-
.28294632730294451032613534596558*i
  .36979370624989001773301777958373-
1.0618231195136687011590573146631*i
  1.1146270688593392975541180081115-
1.36934983263003820407393290085074*i


y =

.43988651431894061086040911013399+1.73489563842
45835120290450101938*i
 -.11319578028438869117106079077457-
1.7344087640028938716691238072308*i
  .17330926596544808031065168066407-
1.21173961516022589100765017648134*i


.17330926596544808031065168066407+1.211739615160
22589100765017648134*i
  -
.11319578028438869117106079077457+1.73440876400
28938716691238072308*i
  .43988651431894061086040911013399-
1.73489563842458351202904501101938*i
```

It is clear from the above that we obtain six complex solutions for the above nonlinear system. Finally, one may use the MATLAB command `format short` to display the six solutions above using four decimal digits only.

In the next chapter, we will present an introduction to calculus using MATLAB.

Exercises

Solve all the exercises using MATLAB. All the needed MATLAB commands for these exercises were presented in this chapter. Note that Exercises 7-11 require the use of the MATLAB Symbolic Math Toolbox.

1. Solve the following linear algebraic equation for the variable x. Use the `roots` command.

$$3x + 5 = 0$$

2. Solve the following quadratic algebraic equation for the variable x. Use the `roots` command.

$$x^2 + x + 1 = 0$$

3. Solve the following algebraic equation for x. Use the `roots` command.

$$3x^4 - 2x^3 + x - 3 = 0$$

4. Solve the following system of linear simultaneous algebraic equations for the variables x and y. Use the inverse matrix method.

$$3x + 5y = 9$$
$$4x - 7y = 13$$

5. In Exercise 4 above, solve the same linear system again but using the Gaussian elimination method with the backslash operator.

6. Solve the following system of linear simultaneous algebraic equations for the variables x, y, and z. Use Gaussian elimination with the backslash operator.

$$2x - y + 3z = 5$$
$$4x + 5z = 12$$
$$x + y + 2z = -3$$

7. Solve the following linear algebraic equation for the variable x in terms of the constant a.

$$2x + a = 5$$

8. Solve the following quadratic algebraic equation for the variable x in terms of the constants a and b.

$$x^2 + ax + b = 0$$

9. Solve the following nonlinear equation for the variable x.

$$2e^x + 3\cos x = 0$$

10. Solve the following system of linear simultaneous algebraic equations for the variables x and y in terms of the constant c.

$$2x - 3cy = 5$$
$$cx + 2y = 7$$

11. Solve the following system of nonlinear simultaneous algebraic equations for the variables x and y.

$$3x^2 - 2x + y = 7$$
$$xy + x = 5$$

Solutions to the Exercises

1. Solve the following linear algebraic equation for the variable x. Use the roots command.

$$3x + 5 = 0$$

```
>> p = [3 5]

p =

      3       5

>> roots(p)

ans =

   -1.6667
```

2. Solve the following quadratic algebraic equation for the variable x. Use the roots command.

$$x^2 + x + 1 = 0$$

```
>> p = [1 1 1]

p =

     1      1      1

>> roots(p)

ans =
```

```
-0.5000 + 0.8660i
-0.5000 - 0.8660i
```

3. Solve the following algebraic equation for x. Use the roots command.

$$3x^4 - 2x^3 + x - 3 = 0$$

```
>> p = [3 -2 0 1 -3]

p =

     3    -2     0     1    -3

>> roots(p)

ans =

    1.1170
    0.2426 + 0.9477i
    0.2426 - 0.9477i
   -0.9355
```

4. Solve the following system of linear simultaneous algebraic equations for the variables x and y. Use the inverse matrix method.

$$3x + 5y = 9$$
$$4x - 7y = 13$$

```
>> A = [3 5 ; 4 -7]

A =

     3     5
     4    -7

>> b = [9 ; 13]
```

```
b =

     9
    13

>> x = inv(A)*b

x =

    3.1220
   -0.0732
```

5. In Exercise 4 above, solve the same linear system again but using the Gaussian elimination method with the backslash operator.

```
>> A = [3 5 ; 4 -7]

A =

    3        5
    4       -7

>> b = [9 ; 13]

b =

     9
    13

>> x = A\b

x =

    3.1220
   -0.0732
```

6. Solve the following system of linear simultaneous algebraic equations for the variables x, y, and z. Use Gaussian elimination with the backslash operator.

$$2x - y + 3z = 5$$
$$4x + 5z = 12$$
$$x + y + 2z = -3$$

```
>> A = [2 -1 3 ; 4 0 5 ; 1 1 2]

A =

     2     -1      3
     4      0      5
     1      1      2

>> b = [5 ; 12 ; -3]

b =

      5
     12
     -3

>> x = A\b

x =

    10.0000
    -1.8000
    -5.6000
```

7. Solve the following linear algebraic equation for the variable
   x in terms of the constant a.

$$2x + a = 5$$

```
>> syms x
>> syms a
>> solve('2*x + a - 5 = 0')

ans =
```

```
-1/2*a+5/2
```

8. Solve the following quadratic algebraic equation for the variable x in terms of the constants a and b.

$$x^2 + ax + b = 0$$

```
>> syms x
>> syms a
>> syms b
>> solve('x^2 + a*x + b = 0')

ans =

 -1/2*a+1/2*(a^2-4*b)^(1/2)
 -1/2*a-1/2*(a^2-4*b)^(1/2)
```

9. Solve the following nonlinear equation for the variable x.

$$2e^x + 3\cos x = 0$$

```
>> syms x
>> solve('2*exp(x) + 3*cos(x) = 0')

ans =

-1.6936679094546100269188394124169
```

The above answer is approximated as -1.694.

10. Solve the following system of linear simultaneous algebraic equations for the variables x and y in terms of the constant c.

$$2x - 3cy = 5$$
$$cx + 2y = 7$$

```
>> syms x
>> syms y
```

```
>> syms c
>> [x,y] = solve('2*x - 3*c*y - 5 = 0',
      'c*x + 2*y - 7 = 0')

x =

(21*c+10)/(4+3*c^2)

y =

-1/(4+3*c^2)*(-14+5*c)
```

11. Solve the following system of nonlinear simultaneous algebraic equations for the variables x and y.

$$3x^2 - 2x + y = 7$$
$$xy + x = 5$$

```
>> syms x
>> syms y
>> [x,y] = solve('3*x^2 - 2*x + y - 7
      = 0', 'x*y + x - 5 = 0')

x =

                5/3
   1/2*5^(1/2)-1/2
  -1/2-1/2*5^(1/2)

y =

                 2
   3/2+5/2*5^(1/2)
3/2-5/2*5^(1/2)
```

4. Differential Equations with MATLAB

The analytical closed-from solutions of ordinary differential equations[9] are presented in this chapter[10]. Details about their solution using the Symbolic Math Toolbox are shown here.

Ordinary differential equations are classified according to the order of the derivative that appears in the equations. First-Order Ordinary Differential Equations have the highest derivative appearing in the equation of the first order. Second-Order Ordinary Differential Equations have the highest derivative appearing in the equation of the second order, and so on. In addition to solving single ordinary differential equations, this chapter also shows how to solve systems of ordinary differential equations using MATLAB.

First-Order Ordinary Differential Equations

In this section we deal with first order differential equations which are of the general form:

$$\frac{dy}{dx} = f(x, y)$$

---

[9] Other types of differential equations are called Partial Differential Equations (PDEs) but these are more complicated and are beyond the scope of this book. PDEs involve partial derivates instead of ordinary derivatives.

[10] This article appeared originally as a chapter in the book "MATLAB for Beginners: A Gentle Approach – Second Edition."

The above equation is called a first-order differential ordinary equation because the derivative $\dfrac{dy}{dx}$ appears raised to the first power only. The function f is a function of the two variables x and y. In particular, x is the independent variable and y is the dependent variable.

The solution of the above differential equation is obtained by finding an expression for y in terms of x. We will use the MATLAB Symbolic Math Toolbox in this chaper to solve such differential equations. Using this toolbox will enable us to obtain analytical solutions to these equations without resorting to numerical methods. In particular, we will make extensive use of the MATLAB command dsolve in the solution of ordinary differential equations. In order to use this command, make sure that you have installed both MATLAB and the MATLAB Symbolic Math Toolbox successfully on your computing system.

Let us first solve the following simple first-order differential equation

$$\frac{dy}{dx} = a \quad , \quad \text{where } a \text{ is a constant.}$$

The above equation is solved using the following use of the dsolve command (which is available in the MATLAB Symbolic Math Toolbox):

```
>> dsolve('Dy = a')

ans =

a*t+C1
```

Here are some comments about the above solution that was generated by MATLAB. Note how the differential equation is written between apostrophes in the argument[11] of the dsolve command. Not also the use of the letter D to denote the derivative of the variable. Note also that the solution obtained is written in terms of the independent variable t. In order to use x for the independent variable, we need to write x explicitly in the argument of the command as follows:

```
>> dsolve('Dy = a', 'x')

ans =

a*x+C1
```

Note that the solution is generated successfully by MATLAB as follows:

$$y = ax + C_1$$

where C_1 is the constant of integration. In order to find the value of the constant C_1, we need to know the initial condition. However, the intial condition is not given in the example, so we leave the final result in terms of C_1 as shown above.

Next, we solve the following differential equation:

$$\frac{dy}{dx} = x$$

---

[11] The argument of a command or a function is the expression appearing inside the parentheses.

The following are the needed MATLAB command `dsolve` to get the solution[12]:

```
>> dsolve('Dy = x', 'x')

ans =

1/2*x^2+C1
```

Thus, the solution is correctly obtained as:

$$y = \frac{1}{2}x^2 + C_1$$

Note that the result shown above leaves the constant of integration C_1 undetermined because we are not given an intial condition in this example. Note also that the constant C_1 of this example is not the same as the constant C_1 of the previous example. It is just that MATLAB uses the same notation for the constant of integration in each example.

Next, let us solve the following differential equation:

$$\frac{dy}{dx} = 5x + 4$$

Here is the MTLAB command:

```
>> dsolve('Dy = 5*x+4', 'x')

ans =

5/2*x^2+4*x+C1
```

---

[12] Note that Dy is used for the first derivative, while D2y is used for the second derivative.

The solution is correctly obtained as:

$$y = \frac{5}{2}x^2 + 4x + C_1$$

Let us next solve the more complicated differential equation:

$$\frac{dy}{dx} = y$$

The following are the MATLAB commands to solve this equation:

```
>> dsolve('Dy = y', 'x')

ans =

C1*exp(x)
```

We correctly obtain the exponential function as the solution as follows:

$$y = C_1 e^x$$

Let us find next the solution to the following first-order differential equation[13] that comes with an initial condition:

$$y' + 3y = e^{-x} \quad , \quad y(0) = 0.45$$

---

[13] The derivative is denoted by dy/dx, or equivalently, by y'.

The following are the MATLAB commands to solve the above equation:

```
>> dsolve('Dy +3*y = exp(-x)', 'y(0) =
0.45', 'x')

ans =

(1/2*exp(2*x)-1/20)*exp(-3*x)
```

Note in the above output the use of the initial condition y(0) = 0.45 in the argument[14] of the `dsolve` command. Thus, the solution is obtained as follows:

$$y = \left(\frac{1}{2}e^{2x} - \frac{1}{20} \right)e^{-3x}$$

Note that the result shown above does not include the constant of integration C_1. The reason is that this example comes with an intial condition, and MATLAB used the intial condition to evaluate the constant of integration, and finally to display the final result that accounts for the intial condition. We show below how the above expression satisfies the initial condition $y(0) = 0.45$:

$$y(0) = \left(\frac{1}{2}e^0 - \frac{1}{20} \right)e^0 = \frac{1}{2} - \frac{1}{20} = \frac{9}{20} = 0.45$$

Next, let us find the solution to the following first-order differential equation:

$$y' + 4xy = x \ , \quad y(0) = 0$$

---

[14] The argument of a command is the expression shown between the parentheses.

The following are the MATLAB commands to solve the above equation:

```
>> dsolve('Dy +4*x*y = x', 'y(0) = 0',
'x')

ans =

1/4-1/4*exp(-2*x^2)
```

Note how the intial condition $y(0) = 0$ is used in the `dsolve` command. The solution is obtained as follows:

$$y = \frac{1}{4} - \frac{1}{4}e^{-2x^2}$$

Note that the above result does not include a constant of integration simply because an intial condition is used in the example. It is clear that the above expression satisfies the initial condition $y(0) = 0$. This is shown below:

$$y(0) = \frac{1}{4} - \frac{1}{4}e^0 = \frac{1}{4} - \frac{1}{4} = 0$$

Let us plot the above function on a graph to show the behavior of the solution in this example. For this purpose, use is made of the MATLAB command `ezplot`[15] that is specially designed to plot functions such as $y = f(x)$. Here are the MATLAB commands needed to plot the above function. The resulting plot is shown in Figure 1.

```
>> ezplot(ans,[-2 2])
```

---

[15] This is the first use of the MATLAB command `ezplot`. This command is not discussed in Chapter 9 on Graphs.

```
>> hold on
>> xlabel('x')
>> ylabel('y')
```

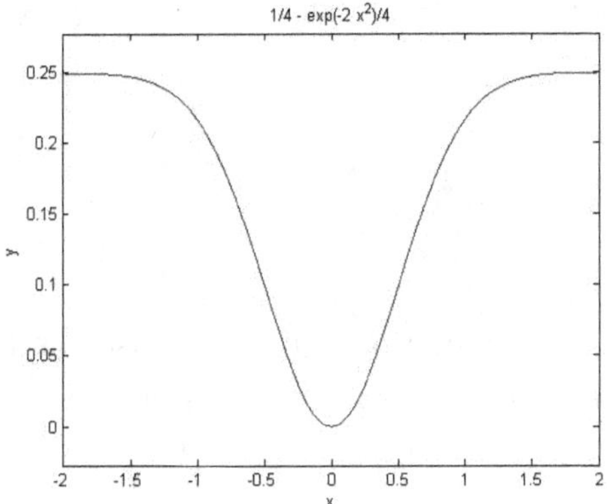

Figure 1: Plot of the Resulting Function in the Previous Example

Next, let us consider the same differential equation as above, but change the initial condition to be in terms of an unspecified constant b, as follows:

$$y' + 4xy = x \ , \quad y(0) = b$$

The MATLAB commands to solve the above differential equation are listed below:

```
>>
dsolve('Dy+4*x*y=x','y(0)=b','x')

ans =

(exp(-2*x^2)*(4*b - 1))/4 + 1/4
```

It is seen from the above MATLAB output that the solution to the differential equation is the following function:

$$y = \frac{1}{4} + \frac{1}{4}(4b - 1)e^{-2x^2}$$

We show next how the above expression satisfies the initial condition $y(0) = b$. Here are the details:

$$y(0) = \frac{1}{4} + \frac{1}{4}(4b - 1)e^0 = \frac{1}{4} + \frac{1}{4}(4b - 1) = b$$

Our next example is to solve the following differential equation:

$$x^2 \frac{dy}{dx} = y \ln y + 2\frac{dy}{dx}$$

The needed MATLAB commands to solve the above equation are shown below. Note the use of the MATLAB command `simplify` on the third command line:

```
>> dsolve('x^2*Dy=y*log(y)+2*Dy','x')

ans =


1
    exp(exp(C11 -  (2^(1/2)*log(x +
2^(1/2)))/4 + (2^(1/2)*log(x -
2^(1/2)))/4))

>> simplify(ans)

ans =


1
    exp((exp(C11)*(x -
2^(1/2))^(2^(1/2)/4))/(x +
2^(1/2))^(2^(1/2)/4))
```

It is seen from the above solution that there is a constant of integration C11. This is clear since there is no initial condition given in this example. The solution is written as follows:

$$y = e^{\dfrac{e^{c11\left(x-\sqrt{2}\right)^{\sqrt{2}/4}}}{\left(x+\sqrt{2}\right)^{\sqrt{2}/4}}}$$

Second-Order Ordinary Differential Equations[16]

In this section we deal with second order differential equations which are of the general form:

$$\frac{d^2y}{dx^2} = f\left(x, y, \frac{dy}{dx}\right)$$

The above equation is called a second-order differential equation because the second derivative $\dfrac{d^2y}{dx^2}$ appears raised to the first power only. The function f is a function of the two variables x and y, as well as the first derivative $\dfrac{dy}{dx}$. In particular, x is the independent variable and y is the dependent variable.

Let us first solve the following second-order ordinary differential equation:

---

[16] There are also higher-order ordinary differential equations of third-order, fourth-order, etc. – check Chapter 17 for solving these types of differential equations.

$$\frac{d^2 y}{dx^2} + 4\frac{dy}{dx} = 2e^{3x} \qquad , \qquad y(0)=1 \ , \ y'(0)=0$$

Note that the above second-order ordinary differential equation comes with two initial conditions; one for y and one for $y' \equiv \dfrac{dy}{dx}$. The following are the MATLAB commands needed to solve the above equation:

```
>>
dsolve('D2y+4*Dy=2*exp(3*x)','y(0)=1','Dy
(0)=0','x')

ans =

exp(-4*x)/14 + exp(-4*x)*((5*exp(4*x))/6
+ (2*exp(7*x))/21)

>> simplify(ans)

ans =

(2*exp(3*x))/21 + exp(-4*x)/14 + 5/6
```

It is noted that the MATLAB command `simplify` is used to simplify the resulting function. Thus, the solution to the above equation is written as follows:

$$y = \frac{5}{6} + \frac{1}{14}e^{-4x} + \frac{2}{21}e^{3x}$$

Let us show that the above expression satisfies the initial condition $y(0)=1$. The details follow:

$$y(0) = \frac{5}{6} + \frac{1}{14}e^{0} + \frac{2}{21}e^{0} = \frac{5}{6} + \frac{1}{14} + \frac{2}{21} = \frac{42}{42} = 1$$

Let us now solve the following second-order ordinary differential equation:

$$5\frac{d^2y}{dx^2}+3\frac{dy}{dx}=2\sin x \quad , \quad y(0)=-1 \ , \ y'(0)=1$$

The MATLAB commands used to solve the above equation are:

```
>> dsolve('5*D2y+3*Dy=2*sin(x)','y(0)=-
1','Dy(0)=1','x')

ans =

4/3 - (3*cos(x))/17 - (5*sin(x))/17 -
(110*exp(-(3*x)/5))/51

>> simplify(ans)

ans =

4/3 - (3*cos(x))/17 - (5*sin(x))/17 -
(110*exp(-(3*x)/5))/51
```

It is noted from the MATLAB output above that the use of the MATLAB command `simplify` did not result in a simplified expression. Thus, the solution to the above differential equation is listed as follows:

$$y=\frac{4}{3}-\frac{3}{17}\cos x-\frac{5}{17}\sin x-\frac{110}{51}e^{-3x/5}$$

Let us check that the above expression satisfies the initial condition $y(0)=-1$. The details follow:

$$y = \frac{4}{3} - \frac{3}{17}\cos 0 - \frac{5}{17}\sin 0 - \frac{110}{51}e^0$$

$$= \frac{4}{3} - \frac{3}{17} - \frac{110}{51} = -\frac{51}{51} = -1$$

Let us now plot the above solution function using the MATLAB command `ezplot`. The needed MATLAB commands are listed below. The resulting plot is shown in Figure 2.

```
>> ezplot(ans)
>> hold on
>> xlabel('x')
>> ylabel('y')
```

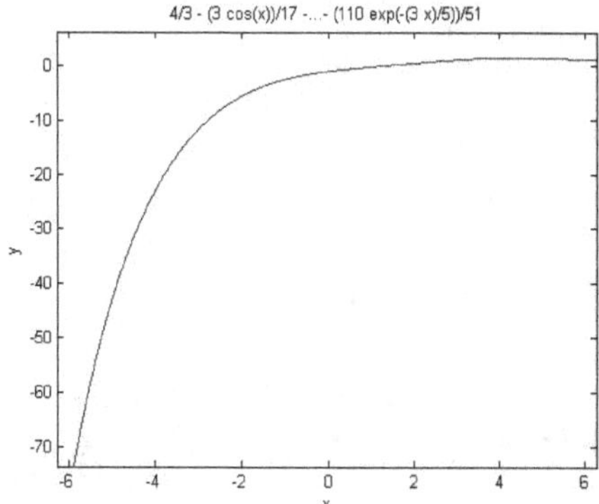

Figure 2: Plot of the Reulting Function of the Previous Example

Systems of Ordinary Differential Equations

In this section we deal with several ordinary differential equations that must be solved simultaneously. Such equations are called systems of ordinary differential equations. We will limit our presentation to linear first-order systems of ordinary differential equations. For this purpose, one example is illustrated.

Let us solve the following linear first-order system of three ordinary differential equations:

$$x'(t) = 2x(t) + 3y(t) - z(t)$$

$$y'(t) = x(t) + 2z(t)$$

$$z'(t) = 4x(t) - 4y(t) + 7z(t)$$

It is noted from the above system of equations that the dependent variables are x, y, and z, while the independent variable is t. It is noted also that all three equations are linear equations. All the derivatives appear on the left-hand-sides of the equations (but this is not a necessary requirement). The above system of equations can be solved using the MATLAB command dsolve as follows:

```
>> [x y z] = dsolve('Dx=2*x+3*y-
z','Dy=x+2*z','Dz=4*x-4*y+7*z')

x =

-(exp(-(t*(3*(4 - (27^(1/2)*368^(1/2))/27)^(2/3) - 18*(4 -
(27^(1/2)*368^(1/2))/27)^(1/3) + 4))/(6*(4 -
(27^(1/2)*368^(1/2))/27)^(1/3))))*(264*C19*cos((t*(4*3^(1/2)
- 3*3^(1/2)*(4 - (27^(1/2)*368^(1/2))/27)^(2/3)))/(6*(4 -
(27^(1/2)*368^(1/2))/27)^(1/3))) -
264*C21*sin((t*(4*3^(1/2) - 3*3^(1/2)*(4 -
(27^(1/2)*368^(1/2))/27)^(2/3)))/(6*(4 -
(27^(1/2)*368^(1/2))/27)^(1/3))) + 72*C19*cos((t*(4*3^(1/2)
- 3*3^(1/2)*(4 - (27^(1/2)*368^(1/2))/27)^(2/3)))/(6*(4 -
(27^(1/2)*368^(1/2))/27)^(1/3)))*(4 -
(27^(1/2)*368^(1/2))/27)^(1/3) + 504*C19*cos((t*(4*3^(1/2)
- 3*3^(1/2)*(4 - (27^(1/2)*368^(1/2))/27)^(2/3)))/(6*(4 -
(27^(1/2)*368^(1/2))/27)^(1/3)))*(4 -
(27^(1/2)*368^(1/2))/27)^(2/3) + 27*C19*cos((t*(4*3^(1/2) -
3*3^(1/2)*(4 - (27^(1/2)*368^(1/2))/27)^(2/3)))/(6*(4 -
(27^(1/2)*368^(1/2))/27)^(1/3)))*(4 -
(27^(1/2)*368^(1/2))/27)^(4/3) - 72*C21*sin((t*(4*3^(1/2) -
```

3*3^(1/2)*(4 - (27^(1/2)*368^(1/2))/27)^(2/3)))/(6*(4 -
(27^(1/2)*368^(1/2))/27)^(1/3)))*(4 -
(27^(1/2)*368^(1/2))/27)^(1/3) - 504*C21*sin((t*(4*3^(1/2)
- 3*3^(1/2)*(4 - (27^(1/2)*368^(1/2))/27)^(2/3)))/(6*(4 -
(27^(1/2)*368^(1/2))/27)^(1/3)))*(4 -
(27^(1/2)*368^(1/2))/27)^(2/3) - 27*C21*sin((t*(4*3^(1/2) -
3*3^(1/2)*(4 - (27^(1/2)*368^(1/2))/27)^(2/3)))/(6*(4 -
(27^(1/2)*368^(1/2))/27)^(1/3)))*(4 -
(27^(1/2)*368^(1/2))/27)^(4/3) - 528*C20*exp((t*(9*(4 -
(27^(1/2)*368^(1/2))/27)^(1/3) + 3*(4 -
(27^(1/2)*368^(1/2))/27)^(2/3) + 4))/(3*(4 -
(27^(1/2)*368^(1/2))/27)^(1/3)))*exp((t*(3*(4 -
(27^(1/2)*368^(1/2))/27)^(2/3) - 18*(4 -
(27^(1/2)*368^(1/2))/27)^(1/3) + 4))/(6*(4 -
(27^(1/2)*368^(1/2))/27)^(1/3))) -
264*3^(1/2)*C21*cos((t*(4*3^(1/2) - 3*3^(1/2)*(4 -
(27^(1/2)*368^(1/2))/27)^(2/3)))/(6*(4 -
(27^(1/2)*368^(1/2))/27)^(1/3))) -
264*3^(1/2)*C19*sin((t*(4*3^(1/2) - 3*3^(1/2)*(4 -
(27^(1/2)*368^(1/2))/27)^(2/3)))/(6*(4 -
(27^(1/2)*368^(1/2))/27)^(1/3))) -
2*27^(1/2)*368^(1/2)*C19*cos((t*(4*3^(1/2) - 3*3^(1/2)*(4 -
(27^(1/2)*368^(1/2))/27)^(2/3)))/(6*(4 -
(27^(1/2)*368^(1/2))/27)^(1/3))) +
2*27^(1/2)*368^(1/2)*C21*sin((t*(4*3^(1/2) - 3*3^(1/2)*(4 -
(27^(1/2)*368^(1/2))/27)^(2/3)))/(6*(4 -
(27^(1/2)*368^(1/2))/27)^(1/3))) - 144*C20*exp((t*(9*(4 -
(27^(1/2)*368^(1/2))/27)^(1/3) + 3*(4 -
(27^(1/2)*368^(1/2))/27)^(2/3) + 4))/(3*(4 -
(27^(1/2)*368^(1/2))/27)^(1/3)))*exp((t*(3*(4 -
(27^(1/2)*368^(1/2))/27)^(2/3) - 18*(4 -
(27^(1/2)*368^(1/2))/27)^(1/3) + 4))/(6*(4 -
(27^(1/2)*368^(1/2))/27)^(1/3)))*(4 -
(27^(1/2)*368^(1/2))/27)^(1/3) + 504*C20*exp((t*(9*(4 -
(27^(1/2)*368^(1/2))/27)^(1/3) + 3*(4 -
(27^(1/2)*368^(1/2))/27)^(2/3) + 4))/(3*(4 -
(27^(1/2)*368^(1/2))/27)^(1/3)))*exp((t*(3*(4 -
(27^(1/2)*368^(1/2))/27)^(2/3) - 18*(4 -
(27^(1/2)*368^(1/2))/27)^(1/3) + 4))/(6*(4 -
(27^(1/2)*368^(1/2))/27)^(1/3)))*(4 -
(27^(1/2)*368^(1/2))/27)^(2/3) - 54*C20*exp((t*(9*(4 -
(27^(1/2)*368^(1/2))/27)^(1/3) + 3*(4 -
(27^(1/2)*368^(1/2))/27)^(2/3) + 4))/(3*(4 -
(27^(1/2)*368^(1/2))/27)^(1/3)))*exp((t*(3*(4 -
(27^(1/2)*368^(1/2))/27)^(2/3) - 18*(4 -
(27^(1/2)*368^(1/2))/27)^(1/3) + 4))/(6*(4 -
(27^(1/2)*368^(1/2))/27)^(1/3)))*(4 -
(27^(1/2)*368^(1/2))/27)^(4/3) +
72*3^(1/2)*C21*cos((t*(4*3^(1/2) - 3*3^(1/2)*(4 -
(27^(1/2)*368^(1/2))/27)^(2/3)))/(6*(4 -
(27^(1/2)*368^(1/2))/27)^(1/3)))*(4 -
(27^(1/2)*368^(1/2))/27)^(1/3) +
27*3^(1/2)*C21*cos((t*(4*3^(1/2) - 3*3^(1/2)*(4 -
(27^(1/2)*368^(1/2))/27)^(2/3)))/(6*(4 -
(27^(1/2)*368^(1/2))/27)^(1/3)))*(4 -

```
(27^(1/2)*368^(1/2))/27)^(4/3) +
72*3^(1/2)*C19*sin((t*(4*3^(1/2) - 3*3^(1/2)*(4 -
(27^(1/2)*368^(1/2))/27)^(2/3)))/(6*(4 -
(27^(1/2)*368^(1/2))/27)^(1/3)))*(4 -
(27^(1/2)*368^(1/2))/27)^(1/3) +
27*3^(1/2)*C19*sin((t*(4*3^(1/2) - 3*3^(1/2)*(4 -
(27^(1/2)*368^(1/2))/27)^(2/3)))/(6*(4 -
(27^(1/2)*368^(1/2))/27)^(1/3)))*(4 -
(27^(1/2)*368^(1/2))/27)^(4/3) +
4*27^(1/2)*368^(1/2)*C20*exp((t*(9*(4 -
(27^(1/2)*368^(1/2))/27)^(1/3) + 3*(4 -
(27^(1/2)*368^(1/2))/27)^(2/3) + 4))/(3*(4 -
(27^(1/2)*368^(1/2))/27)^(1/3)))*exp((t*(3*(4 -
(27^(1/2)*368^(1/2))/27)^(2/3) - 18*(4 -
(27^(1/2)*368^(1/2))/27)^(1/3) + 4))/(6*(4 -
(27^(1/2)*368^(1/2))/27)^(1/3))) +
2*3^(1/2)*27^(1/2)*368^(1/2)*C21*cos((t*(4*3^(1/2) -
3*3^(1/2)*(4 - (27^(1/2)*368^(1/2))/27)^(2/3)))/(6*(4 -
(27^(1/2)*368^(1/2))/27)^(1/3))) +
2*3^(1/2)*27^(1/2)*368^(1/2)*C19*sin((t*(4*3^(1/2) -
3*3^(1/2)*(4 - (27^(1/2)*368^(1/2))/27)^(2/3)))/(6*(4 -
(27^(1/2)*368^(1/2))/27)^(1/3)))))/(864*(4 -
(27^(1/2)*368^(1/2))/27)^(2/3))


y =

(exp(-(t*(3*(4 - (27^(1/2)*368^(1/2))/27)^(2/3) - 18*(4 -
(27^(1/2)*368^(1/2))/27)^(1/3) + 4))/(6*(4 -
(27^(1/2)*368^(1/2))/27)^(1/3)))*(168*C19*cos((t*(4*3^(1/2)
- 3*3^(1/2)*(4 - (27^(1/2)*368^(1/2))/27)^(2/3)))/(6*(4 -
(27^(1/2)*368^(1/2))/27)^(1/3))) -
168*C21*sin((t*(4*3^(1/2) - 3*3^(1/2)*(4 -
(27^(1/2)*368^(1/2))/27)^(2/3)))/(6*(4 -
(27^(1/2)*368^(1/2))/27)^(1/3))) + 72*C19*cos((t*(4*3^(1/2)
- 3*3^(1/2)*(4 - (27^(1/2)*368^(1/2))/27)^(2/3)))/(6*(4 -
(27^(1/2)*368^(1/2))/27)^(1/3)))*(4 -
(27^(1/2)*368^(1/2))/27)^(1/3) + 360*C19*cos((t*(4*3^(1/2)
- 3*3^(1/2)*(4 - (27^(1/2)*368^(1/2))/27)^(2/3)))/(6*(4 -
(27^(1/2)*368^(1/2))/27)^(1/3)))*(4 -
(27^(1/2)*368^(1/2))/27)^(2/3) - 27*C19*cos((t*(4*3^(1/2) -
3*3^(1/2)*(4 - (27^(1/2)*368^(1/2))/27)^(2/3)))/(6*(4 -
(27^(1/2)*368^(1/2))/27)^(1/3)))*(4 -
(27^(1/2)*368^(1/2))/27)^(4/3) - 72*C21*sin((t*(4*3^(1/2) -
3*3^(1/2)*(4 - (27^(1/2)*368^(1/2))/27)^(2/3)))/(6*(4 -
(27^(1/2)*368^(1/2))/27)^(1/3)))*(4 -
(27^(1/2)*368^(1/2))/27)^(1/3) - 360*C21*sin((t*(4*3^(1/2)
- 3*3^(1/2)*(4 - (27^(1/2)*368^(1/2))/27)^(2/3)))/(6*(4 -
(27^(1/2)*368^(1/2))/27)^(1/3)))*(4 -
(27^(1/2)*368^(1/2))/27)^(2/3) + 27*C21*sin((t*(4*3^(1/2) -
3*3^(1/2)*(4 - (27^(1/2)*368^(1/2))/27)^(2/3)))/(6*(4 -
(27^(1/2)*368^(1/2))/27)^(1/3)))*(4 -
(27^(1/2)*368^(1/2))/27)^(4/3) - 336*C20*exp((t*(9*(4 -
(27^(1/2)*368^(1/2))/27)^(1/3) + 3*(4 -
(27^(1/2)*368^(1/2))/27)^(2/3) + 4))/(3*(4 -
```

```
(27^(1/2)*368^(1/2))/27)^(1/3)))*exp((t*(3*(4 -
(27^(1/2)*368^(1/2))/27)^(2/3) - 18*(4 -
(27^(1/2)*368^(1/2))/27)^(1/3) + 4))/(6*(4 -
(27^(1/2)*368^(1/2))/27)^(1/3))) -
168*3^(1/2)*C21*cos((t*(4*3^(1/2) - 3*3^(1/2)*(4 -
(27^(1/2)*368^(1/2))/27)^(2/3)))/(6*(4 -
(27^(1/2)*368^(1/2))/27)^(1/3))) -
168*3^(1/2)*C19*sin((t*(4*3^(1/2) - 3*3^(1/2)*(4 -
(27^(1/2)*368^(1/2))/27)^(2/3)))/(6*(4 -
(27^(1/2)*368^(1/2))/27)^(1/3))) -
2*27^(1/2)*368^(1/2)*C19*cos((t*(4*3^(1/2) - 3*3^(1/2)*(4 -
(27^(1/2)*368^(1/2))/27)^(2/3)))/(6*(4 -
(27^(1/2)*368^(1/2))/27)^(1/3))) +
2*27^(1/2)*368^(1/2)*C21*sin((t*(4*3^(1/2) - 3*3^(1/2)*(4 -
(27^(1/2)*368^(1/2))/27)^(2/3)))/(6*(4 -
(27^(1/2)*368^(1/2))/27)^(1/3))) - 144*C20*exp((t*(9*(4 -
(27^(1/2)*368^(1/2))/27)^(1/3) + 3*(4 -
(27^(1/2)*368^(1/2))/27)^(2/3) + 4))/(3*(4 -
(27^(1/2)*368^(1/2))/27)^(1/3)))*exp((t*(3*(4 -
(27^(1/2)*368^(1/2))/27)^(2/3) - 18*(4 -
(27^(1/2)*368^(1/2))/27)^(1/3) + 4))/(6*(4 -
(27^(1/2)*368^(1/2))/27)^(1/3)))*(4 -
(27^(1/2)*368^(1/2))/27)^(1/3) + 360*C20*exp((t*(9*(4 -
(27^(1/2)*368^(1/2))/27)^(1/3) + 3*(4 -
(27^(1/2)*368^(1/2))/27)^(2/3) + 4))/(3*(4 -
(27^(1/2)*368^(1/2))/27)^(1/3)))*exp((t*(3*(4 -
(27^(1/2)*368^(1/2))/27)^(2/3) - 18*(4 -
(27^(1/2)*368^(1/2))/27)^(1/3) + 4))/(6*(4 -
(27^(1/2)*368^(1/2))/27)^(1/3)))*(4 -
(27^(1/2)*368^(1/2))/27)^(2/3) + 54*C20*exp((t*(9*(4 -
(27^(1/2)*368^(1/2))/27)^(1/3) + 3*(4 -
(27^(1/2)*368^(1/2))/27)^(2/3) + 4))/(3*(4 -
(27^(1/2)*368^(1/2))/27)^(1/3)))*exp((t*(3*(4 -
(27^(1/2)*368^(1/2))/27)^(2/3) - 18*(4 -
(27^(1/2)*368^(1/2))/27)^(1/3) + 4))/(6*(4 -
(27^(1/2)*368^(1/2))/27)^(1/3)))*(4 -
(27^(1/2)*368^(1/2))/27)^(4/3) +
72*3^(1/2)*C21*cos((t*(4*3^(1/2) - 3*3^(1/2)*(4 -
(27^(1/2)*368^(1/2))/27)^(2/3)))/(6*(4 -
(27^(1/2)*368^(1/2))/27)^(1/3)))*(4 -
(27^(1/2)*368^(1/2))/27)^(1/3) -
27*3^(1/2)*C21*cos((t*(4*3^(1/2) - 3*3^(1/2)*(4 -
(27^(1/2)*368^(1/2))/27)^(2/3)))/(6*(4 -
(27^(1/2)*368^(1/2))/27)^(1/3)))*(4 -
(27^(1/2)*368^(1/2))/27)^(4/3) +
72*3^(1/2)*C19*sin((t*(4*3^(1/2) - 3*3^(1/2)*(4 -
(27^(1/2)*368^(1/2))/27)^(2/3)))/(6*(4 -
(27^(1/2)*368^(1/2))/27)^(1/3)))*(4 -
(27^(1/2)*368^(1/2))/27)^(1/3) -
27*3^(1/2)*C19*sin((t*(4*3^(1/2) - 3*3^(1/2)*(4 -
(27^(1/2)*368^(1/2))/27)^(2/3)))/(6*(4 -
(27^(1/2)*368^(1/2))/27)^(1/3)))*(4 -
(27^(1/2)*368^(1/2))/27)^(4/3) +
4*27^(1/2)*368^(1/2)*C20*exp((t*(9*(4 -
(27^(1/2)*368^(1/2))/27)^(1/3) + 3*(4 -
```

```
(27^(1/2)*368^(1/2))/27)^(2/3) + 4))/(3*(4 -
(27^(1/2)*368^(1/2))/27)^(1/3)))*exp((t*(3*(4 -
(27^(1/2)*368^(1/2))/27)^(2/3) - 18*(4 -
(27^(1/2)*368^(1/2))/27)^(1/3) + 4))/(6*(4 -
(27^(1/2)*368^(1/2))/27)^(1/3))) +
2*3^(1/2)*27^(1/2)*368^(1/2)*C21*cos((t*(4*3^(1/2) -
3*3^(1/2)*(4 - (27^(1/2)*368^(1/2))/27)^(2/3)))/(6*(4 -
(27^(1/2)*368^(1/2))/27)^(1/3))) +
2*3^(1/2)*27^(1/2)*368^(1/2)*C19*sin((t*(4*3^(1/2) -
3*3^(1/2)*(4 - (27^(1/2)*368^(1/2))/27)^(2/3)))/(6*(4 -
(27^(1/2)*368^(1/2))/27)^(1/3)))))/(864*(4 -
(27^(1/2)*368^(1/2))/27)^(2/3))


z =

exp(-(t*(3*(4 - (27^(1/2)*368^(1/2))/27)^(2/3) - 18*(4 -
(27^(1/2)*368^(1/2))/27)^(1/3) + 4))/(6*(4 -
(27^(1/2)*368^(1/2))/27)^(1/3)))*(C19*cos((t*(4*3^(1/2) -
3*3^(1/2)*(4 - (27^(1/2)*368^(1/2))/27)^(2/3)))/(6*(4 -
(27^(1/2)*368^(1/2))/27)^(1/3))) - C21*sin((t*(4*3^(1/2) -
3*3^(1/2)*(4 - (27^(1/2)*368^(1/2))/27)^(2/3)))/(6*(4 -
(27^(1/2)*368^(1/2))/27)^(1/3))) + C20*exp((t*(9*(4 -
(27^(1/2)*368^(1/2))/27)^(1/3) + 3*(4 -
(27^(1/2)*368^(1/2))/27)^(2/3) + 4))/(3*(4 -
(27^(1/2)*368^(1/2))/27)^(1/3)))*exp((t*(3*(4 -
(27^(1/2)*368^(1/2))/27)^(2/3) - 18*(4 -
(27^(1/2)*368^(1/2))/27)^(1/3) + 4))/(6*(4 -
(27^(1/2)*368^(1/2))/27)^(1/3))))
```

It is seen from the above MATLAB output that complicated and very long expressions are obtained for the three functions $x(t)$, $y(t)$, and $z(t)$. No attempt is made here to simplify the above expressions, write them out explicitly, or even plot them on a graph. The interested reader may wish to do so with much care.

Exercises

Solve all the exercises using MATLAB. All the needed MATLAB commands for these exercises were presented in this chapter. Note that all of these exercises (1-12) require the use of the MATLAB Symbolic Math Toolbox.

1. Solve the first-order ordinary differential equation:

$$\frac{dy}{dx} = b$$

2. Solve the first-order ordinary differential equation:

$$\frac{dy}{dx} = 3x$$

3. Solve the first-order linear ordinary differential equation with an initial condition:

$$\frac{dy}{dx} = 2x - 3 \qquad , \qquad y(0) = 1$$

4. Plot the solution function obtained in Exercise # 3 above.

5. Solve the first-order linear ordinary differential equation with an initial condition involving an unspecified constant b, given as follows:

$$\frac{dy}{dx} = x + 2y \qquad , \qquad y(0) = b$$

6. Solve the following ordinary differential equation:

$$y' - 2y = \sin(3x)$$

7. Solve the following ordinary differential equation:

$$y' + 3xy = 2x - y$$

8. Solve the following ordinary differential equation:

$$x^2 \frac{dy}{dx} = 2y \ln y$$

9. Solve the following second-order ordinary differential equation with two initial conditions:

$$2\frac{d^2y}{dx^2} + 3\frac{dy}{dx} = e^x \quad , \quad y(0)=1 \quad , \quad y'(0)=1$$

10. Solve the following second-order ordinary differential equation with two initial conditions:

$$\frac{d^2y}{dx^2} - 2\frac{dy}{dx} = \sin(2x) \quad , \quad y(0)=0 \quad , \quad y'(0)=1$$

11. Plot the solution function obtained in Exercise # 10 above.

12. Solve the following system of two linear first-order ordinary differential equations with two initial conditions:

$$x'(t) = 3x(t) - y(t)$$

$$y'(t) = x(t) + y(t)$$

The initial conditions are $x(0)=1$ and $y(0)=1$.

Solutions to the Exercises

```
1.  >> dsolve('Dy = b','x')
```

```
ans =

C2 + b*x
```

The resulting function is $y = bx + C_2$. If you do not specify x as the independent variable in the argument of the command above, MATLAB will use t by default.

2.
```
>> dsolve('Dy = 3*x','x')

ans =

(3*x^2)/2 + C4
```

The resulting function is $y = \dfrac{3}{2}x^2 + C4$.

3.
```
>> dsolve('Dy=2*x-3','y(0)=1','x')

ans =

x*(x - 3) + 1
```

The resulting function is $y = x(x-3)+1$.

Let us show that the above expression satisfies the initial condition $y(0) = 1$. The details follow:

$$y(0) = (0)(0-3)+1 = 0+1 = 1$$

4.
```
>> ezplot(ans)
>> hold on
>> xlabel('x')
>> ylabel('y')
```

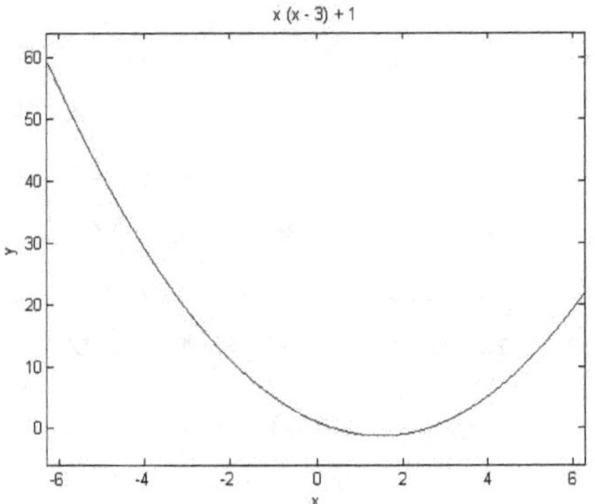

5.  ```
    >> dsolve('Dy=x+2*y','y(0)=b','x')

    ans =

    (exp(2*x)*(4*b + 1))/4 - x/2 - 1/4
    ```

    The resulting function is

    $$y = \frac{1}{4}(4b+1)e^{2x} - \frac{1}{2}x - \frac{1}{4}.$$

    Let us show that the above expression satisties the initial condition $y(0) = b$. The details follow:

    $$y(0) = \frac{1}{4}(4b+1)e^{0} - \frac{1}{2}(0) - \frac{1}{4} = \frac{1}{4}(4b+1) - \frac{1}{4} = \frac{4b}{4} = b$$

6.  ```
    >> dsolve('Dy-2*y=sin(3*x)','x')

    ans =

    C32*exp(2*x) - (2*sin(3*x))/13 -
    (3*cos(3*x))/13
    ```

    The resulting function is

    $$y = C_{32}e^{2x} - \frac{2}{13}\sin(3x) - \frac{3}{13}\cos(3x).$$

7.
```
>> dsolve('Dy+3*x*y=2*x-y','x')

ans =

C34*exp(-(x*(3*x + 2))/2) + exp(-(x*(3*x +
2))/2)*(((-3)^(3/2)*3^(1/2)*exp((3*x^2)/2 +
x)*2*i)/27 + (2^(1/2)*3^(1/2)*pi^(1/2)*exp(-
1/6)*erf((2^(1/2)*3^(1/2)*(3*x + 1)*i)/6)*i)/9)

>> simplify(ans)

ans =

C34*exp(-(x*(3*x + 2))/2) +
(6^(1/2)*pi^(1/2)*erf(6^(1/2)*(x/2 + 1/6)*i)*exp(-
1/6)*exp(-(x*(3*x + 2))/2)*i)/9 + 2/3
```

Note the use of the MATLAB command `simplify` above. The final resulting function is long and involves complicated terms. It will not be shown explicity here, except as presented in the MATLAB output above (Note that `erf` represents the *Error Function*[17], `i`[18] represents an imaginary number, and `pi`[19] is the ratio of the circumference of a circle to its diameter).

8.
```
>> dsolve('x^2*Dy = 2*y*log(y)','x')

ans =

        1
exp(exp(C7 - 2/x))
```

The resulting function is $y = e^{e^{\left(c7 - \frac{2}{x}\right)}}$, where C7 is the constant of integration..

9.
```
>> dsolve('2*D2y+3*Dy=exp(x)','y(0)=1','Dy(0
)=1','x')
```

---

[17] The *Error Function* is one of the special functions in mathematics.
[18] The imaginary number i is equal to the square root of -1.
[19] The value of pi is equal to 3.14….. (never ending decimals) and is approximately equal to 22/7.

```
ans =

exp(x)/5 - (8*exp(-(3*x)/2))/15 + 4/3
```

The resuling function is $y = \dfrac{1}{5}e^x - \dfrac{8}{15}e^{-3x/2} + \dfrac{4}{3}$.

Let is show that the above expression satisfies the initial condition $y(0) = 1$. The details follow:

$$y(0) = \frac{1}{5}e^0 - \frac{8}{15}e^0 + \frac{4}{3} = \frac{1}{5} - \frac{8}{15} + \frac{4}{3} = \frac{15}{15} = 1$$

10.
```
>> dsolve('D2y-
2*Dy=sin(2*x)','y(0)=0','Dy(0)=1','x')

ans =

(5*exp(2*x))/8 - exp(2*x)*((3*exp(-2*x))/4 -
(cos(2*x)*exp(-2*x))/8 + (sin(2*x)*exp(-2*x))/8)

>> simplify(ans)

ans =

(5*exp(2*x))/8 - sin(2*x)/8 + cos(x)^2/4 - 7/8
```

Note the use of the MATLAB command `simplify` above. The resulting function is

$$y = \frac{5}{8}e^{2x} - \frac{1}{8}\sin(2x) + \frac{1}{4}\cos^2 x - \frac{7}{8}.$$

Let us show that the above expression satisfies the initial condition $y(0) = 0$. The details follow:

$$y = \frac{5}{8}e^0 - \frac{1}{8}\sin(0) + \frac{1}{4}\cos^2(0) - \frac{7}{8} = \frac{5}{8} + \frac{1}{4} - \frac{7}{8} = 0$$

11.
```
>> ezplot(ans)
>> hold on
>> xlabel('x')
>> ylabel('y')
```

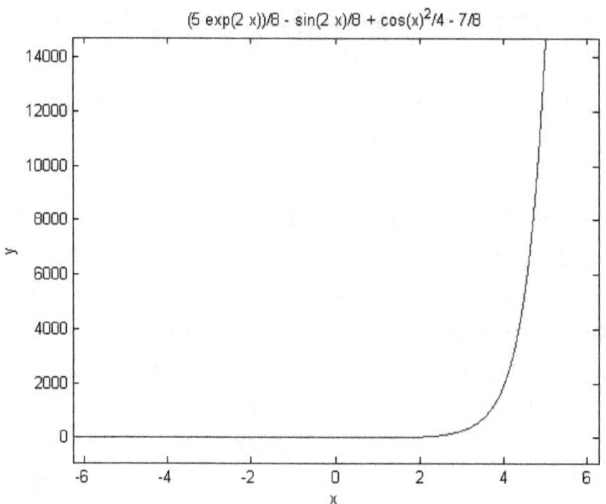

```
12.       >> [x y] = dsolve('Dx=3*x-
   y','Dy=x+y','x(0)=1','y(0)=1','t')

   x =


   exp(2*t)


   y =

   exp(2*t)
```

It is seen from the MATLAB output above that the two solution functions are given by $x(t)=e^{2t}$ and $y(t)=e^{2t}$. Clearly, these two functions satisfy the two initial conditions, as well as the two simultaneous differential equations.

5. Fibonacci Numbers with MATLAB

The Fibonacci sequence is a sequence of numbers where each term of the sequence is obtained by adding the previous two terms. In addition, this special sequence starts with the numbers 1 and 1. Thus, the whole Fibonacci sequence can be generated starting with these two numbers and the above special rule. The numbers in this sequence are called Fibonacci numbers. These numbers were first obtained by the Italian mathematician Leonardo of Pisa (also called Leonardo Fibonacci) in the thirteenth century.

Next, we will generate the first few terms of the Fibonacci sequence. We will start with 1 and 1[20]. Adding these two terms brings us to 2 which is the third Fibonacci number. Then we add 2 and 1 to obtain 3 which is the fourth Fibonacci number. Next, we add 3 and 2 to obtain 5 which is the fifth Fibonacci number. Then, we add 5 and 3 to obtain 8 which is the sixth Fibonacci number. This is followed by adding 8 and 5 to obtain 13 which is the seventh Fibonacci number. This process is illustrated interactively in MATLAB as follows:

```
>> 1 + 1

ans =

    2

>> 2 + 1

ans =

    3
```

---

[20] The Fibonacci sequence can also be started with the numbers 0 and 1 instead of 1 and 1. Similar results will be obtained in this case.

```
>> 3 + 2

ans =

     5

>> 5 + 3

ans =

     8

>> 8 + 5

ans =

    13

>> 13 + 8

ans =

    21

>> 21 + 13

ans =

    34

>> 34 + 21

ans =

    55

>> 55 + 34
```

```
ans =

    89

>> 89 + 55

ans =

   144

>> 144 + 89

ans =

   233
```

The sequence of numbers obtained this way is 1, 1, 2, 3, 5, 8, 13, 21, 34, 55, 89, 144, 233, … is called the Fibonacci sequence. More terms can be generated but the above thirteen terms will suffice for now. Obviously, the Fibonacci sequence goes all the way to infinity as the number of terms increase to infinity. The mathematics of the Fibonacci sequence will be discussed later in this book.

We can write a MATLAB function to readily generate the Fibonacci sequence up until the n-th term. This MATLAB function will be called `fibonacci(n)`. The coding for this function is illustrated below:

```
function f=fibonacci(n)
% This function generates the first n
% Fibonacci numbers
f = zeros(n,1);
f(1)=1;
f(2)=2;
for k = 3:n
```

```
      f(k) = f(k-1) + f(k-2);
end
```

The above MATLAB function can be executed many times; each time with a different value for n, depending on how many terms are needed in the Fibonacci sequence. For example, the first seven Fibonacci numbers are generated by substituting 7 for n in the function fibonacci(n) as follows;

```
>> fibonacci(7)

ans =

        1
        2
        3
        5
        8
       13
       21
```

The function fibonacci(n) can be executed again and again as many times as necessary. For example, the following execution of this function generates the first fifteen Fibonacci numbers;

```
>> fibonacci(15)

ans =

        1
        2
        3
        5
        8
       13
```

```
   21
   34
   55
   89
  144
  233
  377
  610
  987
```

Similarly, the first thirty terms of the Fibonacci sequence are generated as follows:

```
>> fibonacci(30)

ans =

          1
          2
          3
          5
          8
         13
         21
         34
         55
         89
        144
        233
        377
        610
        987
       1597
       2584
       4181
       6765
      10946
```

```
   17711
   28657
   46368
   75025
  121393
  196418
  317811
  514229
  832040
 1346269
```

Thus, we see that the Fibonacci numbers increase exponentially as the value of n increases. We will illustrate this property of the Fibonacci sequence graphically next. But first, let us investigate how to display only the desired Fibonacci number in MATLAB. To accomplish this, we store the Fibonacci sequence in a variable, e.g. y while suppressing the output, then we invoke y(n) for the desired nth Fibonacci number. Here is an example to accomplish this trick inMATLAB.

```
>> y = fibonacci(15);
>> y(15)

ans =

    987
```

Thus, it is seen from the above example that the 15th Fibonacci number is 987. The above trick is especially helpful if we do not want to display the complete Fibonacci sequence of numbers. Another alternative is to revise the MATLAB code for the function fibonacci(n) slightly to display only the desired outcome.

Let us write the rule for generating the Fibonacci sequence using mathematical notation. If we the nth term of the Fibonacci sequence by $F(n)$, then $F(n+1)$ would be the next consecutive Fibonacci number. The previous two Fibonacci numbers would be denoted by $F(n-1)$ and $F(n-2)$. Consequently, the nth Fibonacci number can be generated by the following rule:

$$F(n) = F(n-1) + F(n-2)$$

$$F(1) = 1$$

$$F(2) = 1$$

$$n = 3, 4, 5, \ldots$$

To illustrate using the above mathematical notation, let us show an example with $n = 5$. This would generate the fifth Fibonacci number. Substituting in the first equation above $n = 5$, we obtain:

$$F(5) = F(4) + F(3)$$

To compute both $F(4)$ and $F(3)$, we substitute each one in the same equation again (this equation is called a recursive relation). This, we obtain;

$$F(4) = F(3) + F(2)$$

$$F(3) = F(2) + F(1)$$

Using the initial conditions that $F(1) = 1$ and $F(2) = 1$, and substituting in the above relations, we obtain:

$$F(3) = F(2) + F(1) = 1 + 1 = 2$$

$$F(4) = F(3) + F(2) = 2 + 1 = 3$$

$$F(5) = F(4) + F(3) = 3 + 2 = 5$$

Thus, the fifth Fibonacci number is 5.

Let us plot a graph showing the relationship between the number of terms and the corresponding Fibonacci numbers. In this first plot, let us use only the first ten terms of the Fibonacci sequence. First, let us generate the x-axis of the plot as follows:

```
>> x = 1:10

x =

     1     2     3     4     5     6     7     8     9    10
```

The next step would be to generate the y-axis which would be the first ten Fibonacci numbers. These are generated and stored in the variable y as follows:

```
>> y = fibonacci(10)

y =

     1
     2
     3
     5
     8
    13
    21
    34
    55
    89
```

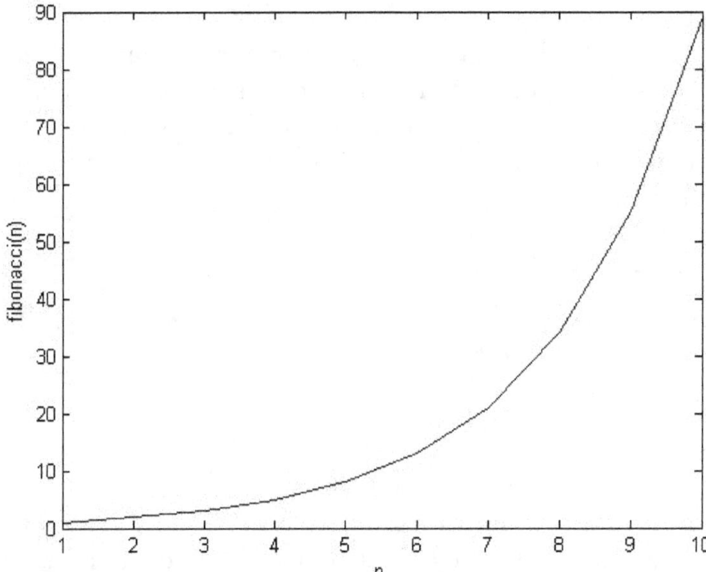

Figure 1: Graph of the first ten Fibonacci numbers

The final step would be to use the MATLAB command `plot` in order to generated the required graph. In addition, we use the MATLAB commands `xlabel` and `ylabel` to label the two axes, respectively. These commands are executed as shown below and the graph is shown in Figure 1.

```
>> plot(x,y)
>> hold on
>> xlabel('n')
>> ylabel('fibonacci(n)')
```

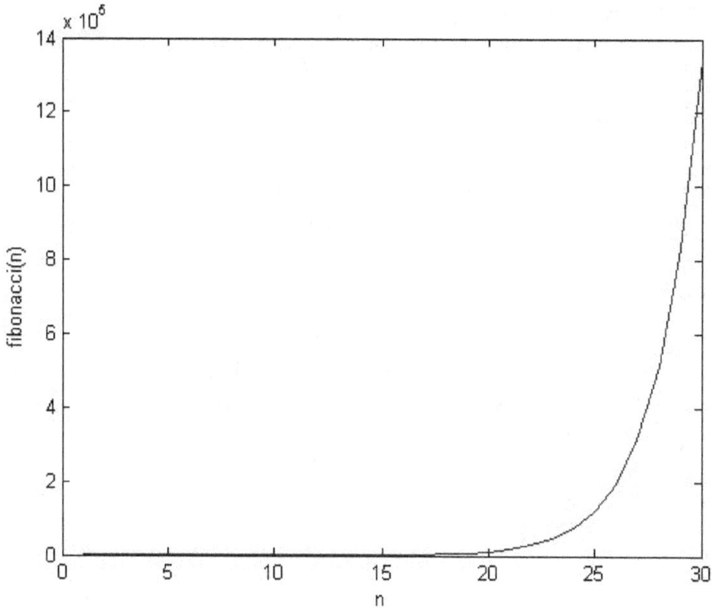

Figure 2: Graph of the first thirty Fibonacci numbers

Similarly, the following commands will generate and plot the first 30 Fibonacci numbers. The plot is generated after the commands are executed sequentially and is shown in Figure 2. Each command is followed by a semi-colon in order to suppress the listed output.

```
>> x = 1:30;
>> y = fibonacci(30);
>> plot(x,y)
>> hold on;
>> xlabel('n')
>> ylabel('fibonacci(n)')
```

It is seen from the above graph that Fibonacci numbers increase exponentially as the value of n increases. In the next chapter, we will investigate the Fibonacci sequence in more details and study its exact property that makes this sequence special – this would be its relationship to what is called the Golden Ratio.

Using the mathematical notation of this chapter, the two graphs in Figures 1 and 2 are called *F(n) vs. n graphs*. Next, we will study the graphs obtained when plotting the logarithms of Fibonacci numbers.

First, let us use the MATLAB function `log` to calculate the logarithms[21] of the first ten Fibonacci numbers as follows:

```
>> log(fibonacci(10))

ans =

        0
   0.6931
   1.0986
   1.6094
   2.0794
   2.5649
   3.0445
   3.5264
   4.0073
   4.4886
```

We need now to plot the above logarithms on the y-axis while keeping the number of terms n on the x-axis. The type of plot obtained is called a semi-log graph. We will perform the following operations (while suppressing the output) in order to accomplish this. The resulting graph is

---

[21] The logarithms calculated here are the natural logarithms, which are logarithms to the base e , where e = 2.71828....

shown in Figure 3. This graph is called a *log(F(n)) vs. n graph*. The MATLAB commands are shown below.

```
>> x = 1:10;
>> y = log(fibonacci(10));
>> plot(x,y)
>> hold on
>> xlabel('n')
>> ylabel('log(F(n))')
```

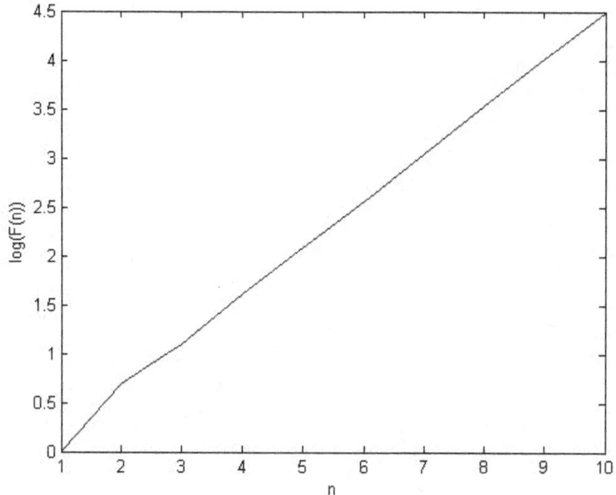

Figure 3: Semi-log graph of the first 10 Fibonacci numbers

It is seen from Figure 3 that the graph is almost a straight line. Let us calculate the slope[22] of this line. Let us perform this calculation by taking the fifth and tenth Fibonacci numbers and calculating the slope of the line segment between them. This is implemented in MATLAB as follows:

---

[22] The slope m of a line segment between two points (x_1, y_1) and (x_2, y_2) is calculated using the formula $m = (y_2 - y_1)/(x_2 - x_1)$.

```
>> m = (y(10)-y(5))/(x(10)-x(5))

m =

    0.4818
```

Thus the slope of the straight line is 0.4818. In order to find the angle, we calculate the inverse tangent value of this number. This is obtained as $\tan^{-1}(0.4818) = 25.7°$.

Let us perform the above operations again but using the first 30 Fibonacci numbers. Here are the MATLAB commands to generate the plot while the graph is shown in Figure 4.

```
>> x = 1:30;
>> y = log(fibonacci(30));
>> plot(x,y)
>> hold on
>> xlabel('n')
>> ylabel('log(F(n))')
```

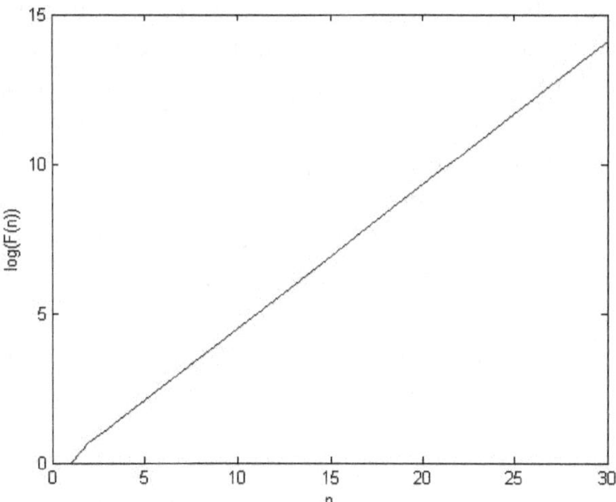

Figure 4: Semi-log graph of the first 30 Fibonacci numbers

As expected, the semi-log graph shows almost a straight line. Let us now calculate the slope of this line between the 10th and the 30th Fibonacci numbers. This operation is implemented in MATLAB as follows:

```
>> m = (y(30)-y(10))/(x(30)-x(10))

m =


   0.4812
```

As expected, the slope of the line is almost as before. The angle of the line is calculated as $\tan^{-1}(0.4812) = 25.7°$, exactly as before. An alternative method to plot the logarithms of Fibonacci numbers is to use the MATLAB command `semilogy` in association with the MATLAB function `fibonacci(n)`. The graph obtained in this way will also be an almost straight line but will be slightly different from the graphs obtained here.

Fibonacci numbers are closely related to exponential growth in nature. Botanists noticed that there are many plants that tend to have a Fibonacci number for the leaves and petals. Fibonacci numbers also appear prominently in the family tree of rabbits. Actually, Leonardo Fibonacci first noticed these numbers when he studied the problem of breeding of rabbits under ideal circumstances. Let us elaborate on this problem and how it is related to Fibonacci numbers. See the books in references [1-9] and the web links [10-20].

Consider a hypothetical situation of rabbits breeding as follows. Start with two rabbits who after a time produce two new rabbits. After a certain time, these four rabbits produce four new rabbits. Thus, we end up with eight rabbits. In the next cycle of breeding, another eight rabbits are produced to make the total 16 rabbits and so on. Thus, we get the exponential sequence 2, 4, 8, 16, 32, ... which is just a sequence of the powers of 2, i.e. 2^1, 2^2, 2^3, 2^4, 2^5, ... This sequence follows the law of exponential growth. See the books in references [1-9] and the web links [10-20].

However, the real situation is different. When we have four rabbits, the newborn rabbits are too young to produce new rabbits right away. They have to skip a cycle before they start producing more rabbits. In this way, the number of rabbits accumulating follows the Fibonacci sequence 1, 1, 2, 3, 5, 8, 13, 21, ... and so on. This is how Leonardo Fibonacci first deduced this interesting series by considering how rabbits breed naturally. This is explained in detail below. See the books in references [1-9] and the web links [10-20].

Let us start with one rabbit. After one cycle, this rabbit produces a new rabbit – the total becomes two. After the second cycle, the old rabbit produces another new rabbit but the little rabbit skips this cycle and does not produce any new rabbits. The total now becomes three rabbits. After another cycle, the two old rabbits produce two new rabbits but the little one skips this cycle and does not produce. Thus the total now becomes five rabbits following the Fibonacci sequence precisely. Thus, we obtain the numbers 1, 2, 3, 5, 8, 13, 21, …. for the total number of rabbits. This is exactly what Leonardo Fibonacci originally observed and is what started the interesting topic of Fibonacci numbers. See the books in references [1-9] and the web links [10-20].

Fibonacci numbers appear in nature often enough to show that they reflect some naturally occurring patterns. These patterns can be spotted by studying how certain plants grow. For example, if you look at an array of seeds in the center of a sunflower, you will notice what looks like spiral patterns curving left and right. If you count these spirals, you will be surprised to obtain a Fibonacci number. If you divide the spirals into those pointed left and right, you will get two consecutive Fibonacci numbers. You can find such spiral patterns in pinecones, pineapples, and cauliflower. See the books in references [1-9] and the web links [10-20].

Other examples where Fibonacci numbers occur naturally are in plant growth and branches of plants. For example, if you count the number of petals on a flower, you will be surprised to find the total to be a Fibonacci number. Lilies and irises have three petals while buttercups and wild roses have five. Delphiniums have eight petals and so on. Notice that 3, 5, and 8 are Fibonacci numbers. See the books in references [1-9] and the web links [10-20].

Fibonacci numbers also occur naturally in honeybee colonies. In these colonies, there is a queen, a few drones, and a lot of workers. The female bees all have two parents – a drone and a queen. However, drones have one parent each. Therefore, Fibonacci numbers appear in a drone's family tree – each drone has one parent, two grandparents, three great-grandparents, and so on. See the books in references [1-9] and the web links [10-20].

A final example is that Fibonacci numbers appear in the human body. You have one nose, two eyes, three segments to each limb, and five fingers on each hand. These are all Fibonacci numbers. Furthermore, DNA molecules follow the Fibonacci sequence also measuring 34 angstroms long and 21 angstroms wide for each full cycle of the double helix. Of course, the numbers 34 and 21 are surprisingly Fibonacci numbers. See the books in references [1-9] and the web links [10-20].

References

Books on Fibonacci Numbers

1. Posamentier, A. S., *The Fabulous Fibonacci Numbers*, Prometheus Books, 2007.

2. Vajda, S., *Fibonacci and Lucas Numbers, and the Golden Section*, Dover Publications, 2007.

3. Vorobiev, N. N. and Martin, M., *Fibonacci Numbers*, Birkhauser, 2003.

4. Dunlap, R. A., *The Golden Ratio and Fibonacci Numbers*, World Scientific Publishing, 1997.

5. Olsen, S., *The Golden Section: Nature's Greatest Secret*, Walker and Company, 2006.

6. Livio, M., *The Golden Ratio: The Story of PHI, the World's Most Astonishing Number*, Broadway, 2003.

7. Koshy, T., *Fibonacci and Lucas Numbers with Applications*, Wiley-Interscience, 2001.

8.   Dobson, E. D., *Understanding Fibonacci Numbers*, Traders Press, 1994.

9.   Hoggatt, V. E. and Bicknell, M., *Primer on Fibonacci Numbers*, Fibonacci Association, 1972.

Web Links for Fibonacci Numbers

10. Fibonacci Numbers at Wolfram MathWorld
http://mathworld.wolfram.com/FibonacciNumber.html

11. Fibonacci Numbers at Wikipedia
http://en.wikipedia.org/wiki/Fibonacci_number

12. Fibonacci Numbers and the Golden Section
http://www.maths.surrey.ac.uk/hosted-sites/R.Knott/Fibonacci/fib.html

13. The Mathematical Magic of Fibonacci Numbers
http://www.maths.surrey.ac.uk/hosted-sites/R.Knott/Fibonacci/fibmaths.html

14. The Fibonacci Sequence – Math is Fun
http://www.mathsisfun.com/numbers/fibonacci-sequence.html

15. Generalizations of Fibonacci Numbers from Wikipedia
http://en.wikipedia.org/wiki/Tribonacci_number

16. Random Fibonacci Sequences from Wikipedia
http://en.wikipedia.org/wiki/Viswanath%27s_constant

17. The Golden Mean and the Physics of Aesthetics by Subhash Kak
http://arxiv.org/abs/physics/0411195

18. Stepping Beyond Fibonacci Numbers – from Ivars Peterson's MathTrek
http://www.maa.org/mathland/mathtrek_09_30_02.html

19. Viswanath's Constant from PlanetMath
http://planetmath.org/encyclopedia/ViswanathsConstant.html

20. Divakar Viswanath, "Random Fibonacci Sequences and the Number 1.13198824…", Mathematics of Computation, Vol. 69, 2000, pp. 1131-1155.

6. The Golden Ratio with MATLAB

This chapter consists of two parts. The first part deals with the basic concepts and equations of the Golden Ratio, while the second part deals with additional and advanced properties of the Golden Ratio.

6.1 Basic Concept and Equations

In this chapter[23], we will continue our study of the Fibonacci sequence and illustrate its precise relationship to what is called the Golden Ratio. Let us calculate the ratio of each two consecutive numbers in the Fibonacci sequence.

Here are the first few terms of the Fibonacci sequence:

1, 1, 2, 3, 5, 8, 13, 21, 34, 55, 89, 144, ...

Let us start by calculating the ratio of the first two terms. The first two terms are 1 and 1; their ratio is $1/1 = 1$. The next two consecutive terms are 1 and 2; their ratio is $2/1 = 2$. Next are the numbers 2 and 3; their ratio is $3/2 = 1.5$. The ratio of 3 and 5 is $5/3 = 1.6667$. Next are the numbers 5 and 8 whose ratio is $8/5 = 1.6$. This followed by the numbers 8 and 13 whose ratio is $13/8 = 1.625$. Next are 13 and 21 whose ratio is $21/13 = 1.6154$. This is followed by the numbers 21 ad 34; their ratio is $34/21 = 1.6190$. Next are the numbers 34 and 55 whose ratio is $55/34 = 1.6176$. The ratio of the next two numbers in the sequence is $89/55 = 1.6182$. Finally, the last two terms in the sequence above have the ratio $144/89 = 1.6180$.

---

[23] This section appeared originally as a chapter in the book "MATLAB Guide to Fibonacci Numbers and the Golden Ratio."

As can be seen from the above calculations, the ratio of any two consecutive Fibonacci numbers seems to approach the number 1.6180 as the numbers increase. In order to ascertain this observation, the above calculations are easily implemented using MATLAB. First, let us perform the above calculations again using MATLAB interactively as follows:

```
>> 1/1

ans =

     1

>> 2/1

ans =

     2

>> 3/2

ans =

    1.5000

>> 5/3

ans =

    1.6667

>> 8/5

ans =

    1.6000
```

```
>> 13/8

ans =

    1.6250

>> 21/13

ans =

    1.6154

>> 34/21

ans =

    1.6190

>> 55/34

ans =

    1.6176

>> 89/55

ans =

    1.6182

>> 144/89

ans =

    1.6180
```

```
>> 233/144

ans =

    1.6181

>> 377/233

ans =

    1.6180

>> 610/377

ans =

    1.6180

>> 987/610

ans =

    1.6180
```

It is clearly seen from the above MATLAB output that the sought ratio approaches the magical number 1.6180. Next, a MATLAB function is shown which calculates this ratio as above and plots the graph of the ratio as the number of terms increases. Here is the code of the MATLAB function `ratio(n)` which calculates this limiting ratio:

```
function r=ratio(n)
% This function calculates the ratio of
two %  consecutive Fibonacci numbers
f = zeros(n,1);
r = zeros(n,1);
f(1)=1;
```

```
f(2)=2;
for k = 3:n
    f(k) = f(k-1) + f(k-2);
    r(k) = f(k)/f(k-1);
end
```

Let us execute the above functions for the first fifteen terms of the Fibonacci sequence. Here is the output of this command:

```
>> ratio(15)

ans =

         0
         0
    1.5000
    1.6667
    1.6000
    1.6250
    1.6154
    1.6190
    1.6176
    1.6182
    1.6180
    1.6181
    1.6180
    1.6180
    1.6180
```

It is seen from the above output that the ratio of two consecutive Fibonacci numbers approaches the value 1.6180. To get more precision, we will use the format long command and execute the function again but with thirty terms as follows:

```
>> format long
```

```
>> ratio(30)

ans =

                         0
                         0
         1.50000000000000
         1.66666666666667
         1.60000000000000
         1.62500000000000
         1.61538461538462
         1.61904761904762
         1.61764705882353
         1.61818181818182
         1.61797752808989
         1.61805555555556
         1.61802575107296
         1.61803713527851
         1.61803278688525
         1.61803444782168
         1.61803381340013
         1.61803405572755
         1.61803396316671
         1.61803399852180
         1.61803398501736
         1.61803399017560
         1.61803398820533
         1.61803398895790
         1.61803398867044
         1.61803398878024
         1.61803398873830
         1.61803398875432
         1.61803398874820
         1.61803398875054
```

It is seen from the above results that the limiting ratio approaches the value 1.6180339887….. (accurate to eleven decimal digits). Let us plot a graph of this ratio with the following commands where the output is suppressed:

```
>> x = 1:30;
>> y = ratio(30);
>> plot(x,y)
>> hold on;
>> xlabel('n');
>> ylabel('F(n)/F(n-1)')
```

The resulting graph is shown in Figure 1. It is seen that the ratio F(n)/F(n-1) approaches the limiting value 1.61803.... . The graph shown in the figure is known as the *F(n)/F(n-1) vs. n* graph.

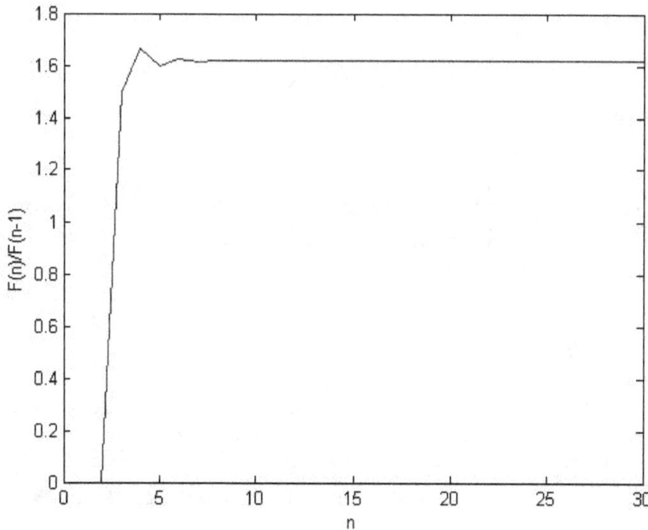

Figure 1: Graph of the limiting ratio of the Fibonacci sequence (shown for the first thirty terms of the sequence). The limiting ration clearly approaches the value 1.61803...

Next, we will investigate the number 1.61803… in more detail. This number is called the Golden Ratio or Golden Section or Divine Proportion. It will be seen that this number arises naturally from certain geometrical constructions. For example, if we divide a straight line into two segments according to a certain rule, we will obtain the Golden Ratio.

Consider a straight line AB which we would like to divide into two segments AC and CB as shown in Figure 2. The rule that we will use to obtain the point C is that the ratio of the length of the straight line to the length of the longer segment is equal to the ratio of the length of the longer segment to the length of the shorter segment. This ratios AB/AC and AC/CB should be equal. Thus, we have the following equation:

Figure 2: Dividing a straight line into two segments in deriving the Golden Ratio

$$\frac{AB}{AC} = \frac{AC}{CB}$$

However, we know that AB = AC + CB. Substituting this relation in the above equation, we obtain:

$$\frac{AC + CB}{AC} = \frac{AC}{CB}$$

Next, we will re-write the above equation as follows:

$$1 + \frac{CB}{AC} = \frac{AC}{CB}$$

Let the ratio AC/CB be denoted by the variable x, then the above equation may be re-written as follows:

$$1 + \frac{1}{x} = x$$

Multiplying both sides of the above equation by x, we obtain the following quadratic equation:

$$x + 1 = x^2$$

Next, we will write the above equation in the standard form of a quadratic equation as follows (by re-arranging the terms):

$$x^2 - x - 1 = 0$$

The above equation can be solved using the quadratic formula but we will use MATLAB to obtain the two solutions. In order to accomplish this, we use the MATLAB command `solve`[24] as follows:

```
>> solve('x^2 - x -1 = 0')

ans =

1/2*5^(1/2)+1/2
1/2-1/2*5^(1/2)
```

---

[24] The `solve` command is part of the MATLAB Symbolic Math Toolbox. You need to have this toolbox installed along with MATLAB in order to use this command.

Thus, it is seen that the two solutions obtained are $\dfrac{1+\sqrt{5}}{2}$ and $\dfrac{1-\sqrt{5}}{2}$. Clearly, the second solution $\dfrac{1-\sqrt{5}}{2}$ is a negative number and should be discarded with regard to our example. Therefore, the positive solution $\dfrac{1+\sqrt{5}}{2}$ obtained is the one that is adopted.

Using MATLAB, we can now easily calculate the numerical value of the positive solution as follows:

```
>> (1+sqrt(5))/2

ans =

    1.6180
```

Thus, it is seen that the solution is exactly the Golden Ratio – the limiting ratio of the Fibonacci sequence. In order to ascertain this, we repeat the above calculation but with more precision as follows:

```
>> format long
>> (1+sqrt(5))/2

ans =

    1.61803398874989
```

Exactly, the same magical number is obtained which is the Golden Ratio. This number is usually denoted by the Greek symbol ϕ. Thus, we have $\phi = 1.618033987$ .... .

Going back to the two solutions of the quadratic equation, let us find the numerical value of the second solution (the negative solution that was discarded) which is $\dfrac{1-\sqrt{5}}{2}$. Using MATLAB, we find its numerical value as follows:

```
>> (1-sqrt(5))/2

ans =

   -0.6180
```

It is seen from the above calculation that the numerical value of the second solution is related to the value of the Golden Ratio. In order to find the exact relationship between the two solutions, let us repeat the above calculation but using more precision as follows:

```
>> format long
>> (1-sqrt(5))/2

ans =

  -0.61803398874989
```

It is seen from the above calculation that the second solution is equal to $1-\phi$. Thus the two solutions of the quadratic equation are ϕ and $1-\phi$. In order to check these that these two values are really the solutions, we substitute each one back into the quadratic equation as follows:

$$\phi^2 - \phi - 1 = 0$$

$$(1-\phi)^2 - (1-\phi) - 1 = 0$$

The above two equations are checked with MATLAB as follows:

```
>> phi = (1+sqrt(5))/2

phi =

    1.6180

>> phi^2 - phi - 1

ans =

     0

>> (1-phi)^2 - (1-phi) - 1

ans =

     0
```

Thus, it is ascertained that both values are solutions of the quadratic equation. We summarize what we have obtained thus far as follows:

$$\phi = \frac{1+\sqrt{5}}{2} = 1.61803398874989$$

$$\phi = \frac{1-\sqrt{5}}{2} = -0.61803398874989$$

In the next chapter, we will investigate certain properties of the Golden Ratio ϕ in more detail.

6.2 Properties of the Golden Ratio

In this chapter[25], we will study certain properties of the Golden Ratio in detail. In addition, we will study a nice formula that is used to compute Fibonacci numbers using the value of the Golden Ratio.

First, let us calculate the inverse of the Golden Ratio ϕ, i.e. let us calculate numerically the value of $\dfrac{1}{\phi}$. This computation is performed in MATLAB as follows:

```
>> phi = (1+sqrt(5))/2

phi =

    1.6180

>> 1/phi

ans =

    0.6180
```

---

[25] This section appeared originally as a chapter in the book "MATLAB Guide to Fibonacci Numbers and the Golden Ratio."

It is seen from the above computation that the value of $\dfrac{1}{\phi}$ is surprisingly equal to $\phi - 1$ which is 0.6180. Thus, we obtain the following formula for ϕ:

$$\frac{1}{\phi} = \phi - 1$$

In order to check the validity of the above formula, we multiply both sides by ϕ to get our familiar quadratic equation $\phi^2 - \phi - 1 = 0$.

Next, let us calculate the square of the Golden Ratio, i.e. the value of ϕ^2. This operation is performed in MATLAB as follows:

```
>> phi = (1+sqrt(5))/2

phi =

    1.6180

>> phi^2

ans =

    2.6180
```

It is seen that another surprising result is obtained. The value of ϕ^2 is seen to be equal to $\phi + 1$ which is 2.6180. Thus, we have the following relation:

$$\phi^2 = \phi + 1$$

Again we obtain the same familiar quadratic equation $\phi^2 - \phi - 1 = 0$.

We have found out that the Golden Ratio has some interesting properties: its reciprocal is equal to $\phi - 1$ and its square is equal to $\phi + 1$. Let us see what other interesting properties this number has.

Let us now study the powers of ϕ, i.e. let us study the terms ϕ^2, ϕ^3, ϕ^4, ... and so on. These powers are given by the following set of equations:

$$\phi^2 = \phi = 1$$

$$\phi^3 = \phi^2 + \phi$$

$$\phi^4 = \phi^3 + \phi^2$$

$$\cdots\cdots\cdots\cdots$$

$$\phi^{n+2} = \phi^{n+1} + \phi^n \quad , \quad n = 0,1,2,3,\ldots$$

We have already checked the second power of ϕ using MATLAB before. Let us now check the third and fourth power of ϕ using MATLAB in order to verify the above equations. Here is the MATLAB output to check the equation for ϕ^3:

```
>> phi = (1+sqrt(5))/2

phi =

    1.6180
```

```
>> phi^3

ans =

    4.2361

>> phi^2 + phi

ans =

    4.2361
```

Now, let us show the MATLAB output to check the equation for ϕ^4 as follows:

```
>> phi = (1+sqrt(5))/2

phi =

    1.6180

>> phi^4

ans =

    6.8541

>> phi^3 + phi^2

ans =

    6.8541
```

Thus, the above equations for the powers of ϕ are verified. There is another set of equations for the powers of ϕ that is also interesting. This new set of equations is written as follows:

$$\phi^2 = \phi + 1$$

$$\phi^3 = 2\phi + 1$$

$$\phi^4 = 3\phi + 2$$

$$\phi^5 = 5\phi + 3$$

$$\phi^6 = 8\phi + 5$$

$$\phi^7 = 13\phi + 8$$

We can see that a pattern emerges for the expressions of the powers of ϕ shown above. It is clear that the numbers appearing in the expressions on the right-hand-side of the equations are Fibonacci numbers. Thus, the above equations serve to strengthen the relationship between Fibonacci numbers and the Golden Ratio. Let us now check the equations for ϕ^3 and ϕ^4 shown above using MATLAB. First, the expression for ϕ^3 is verified as follows:

```
>> phi = (1+sqrt(5))/2

phi =

    1.6180

>> phi^3

ans =

    4.2361

>> 2*phi + 1
```

```
ans =

    4.2361
```

Next, the expression for ϕ^4 is also verified as follows:

```
>> phi = (1+sqrt(5))/2

phi =

    1.6180

>> phi^4

ans =

    6.8541

>> 3*phi +2

ans =

    6.8541
```

Next, we will derive the expressions for the previous two equations for ϕ^3 and ϕ^4 mathematically as follows. Let us start with the expression $\phi^3 = \phi^2 + \phi$ and substitute $\phi + 1$ for ϕ^2 to obtain $\phi^3 = (\phi + 1) + \phi = 2\phi + 1$. Similarly, we start with the expression $\phi^4 = \phi^3 + \phi^2$ and substitute $2\phi + 1$ for ϕ^3, and $\phi + 1$ for ϕ^2 to obtain $\phi^4 = (2\phi + 1) + (\phi + 1) = 3\phi + 2$. We can continue in a similar fashion to prove the other expressions shown above and generate the Fibonacci numbers in this way.

Next, let us study the inverse powers of ϕ, i.e. let us study the expressions for $\dfrac{1}{\phi}, \dfrac{1}{\phi^2}, \dfrac{1}{\phi^3}, \dfrac{1}{\phi^4}, \ldots$ and so on. The following expressions will be shown to be true for the inverse powers of the Golden Ratio:

$$1 = \frac{1}{\phi} + \frac{1}{\phi^2}$$

$$\frac{1}{\phi} = \frac{1}{\phi^2} + \frac{1}{\phi^3}$$

$$\frac{1}{\phi^2} = \frac{1}{\phi^3} + \frac{1}{\phi^4}$$

$$\frac{1}{\phi^3} = \frac{1}{\phi^4} + \frac{1}{\phi^5}$$

$$\ldots\ldots\ldots\ldots$$

$$\frac{1}{\phi^n} = \frac{1}{\phi^{n+1}} + \frac{1}{\phi^{n+2}} \quad , \qquad n = 0,1,2,3,\ldots\ldots$$

We will check now the first two equations above using MATLAB. The first equation $1 = \dfrac{1}{\phi} + \dfrac{1}{\phi^2}$ is checked using MATLAB as follows:

```
>> phi = (1+sqrt(5))/2

phi =
    1.6180

>> 1/phi + 1/phi^2
```

```
ans =

     1
```

Next, the second equation $\dfrac{1}{\phi} = \dfrac{1}{\phi^2} + \dfrac{1}{\phi^3}$ is checked as follows using MATLAB:

```
>> phi = (1+sqrt(5))/2

phi =

    1.6180

>> 1/phi^2 + 1/phi^3

ans =

    0.6180

>> 1/phi

ans =

    0.6180
```

Similarly, the other equations can be verified using MATLAB. There is another set of equations for the inverse powers of ϕ that is interesting. The following equations will be verified to be true:

$$\frac{1}{\phi^2} = 1 - \frac{1}{\phi}$$

$$\frac{1}{\phi^3} = \frac{2}{\phi} - 1$$

$$\frac{1}{\phi^4} = 2 - \frac{3}{\phi}$$

$$\frac{1}{\phi^5} = \frac{5}{\phi} - 3$$

$$\frac{1}{\phi^6} = 5 - \frac{8}{\phi}$$

$$\frac{1}{\phi^7} = \frac{13}{\phi} - 8$$

We can see that a pattern emerges for the expressions of the inverse powers of ϕ shown above. It is clear that the numbers appearing in the expressions on the right-hand-side of the equations are Fibonacci numbers. Thus, the above equations serve to strengthen the relationship between Fibonacci numbers and the Golden Ratio. Let us now check the equations for $\frac{1}{\phi^3}$ and $\frac{1}{\phi^4}$ shown above using MATLAB.

First, the expression for $\frac{1}{\phi^3}$ is verified as follows:

```
>> phi = (1+sqrt(5))/2

phi =

    1.6180

>> 1/phi^3
```

```
ans =

    0.2361

>> 2/phi -1

ans =

    0.2361
```

Next, the expression for $\dfrac{1}{\phi^4}$ is verified using MATLAB as follows:

```
>> phi = (1+sqrt(5))/2

phi =

    1.6180

>> 1/phi^4

ans =

    0.1459

>> 2 - 3/phi

ans =

    0.1459
```

Similarly, the other equations shown above for the inverse powers of ϕ can be verified using MATLAB.

Next, we will derive the expressions for the previous two equations for $\dfrac{1}{\phi^3}$ and $\dfrac{1}{\phi^4}$ mathematically as follows. Let us start with the expression $\dfrac{1}{\phi} = \dfrac{1}{\phi^2} + \dfrac{1}{\phi^3}$ and substitute $1 - \dfrac{1}{\phi}$ for $\dfrac{1}{\phi^2}$ to obtain $\dfrac{1}{\phi^3} = \dfrac{1}{\phi} - \dfrac{1}{\phi^2} = \dfrac{1}{\phi} - \left(1 - \dfrac{1}{\phi}\right) = \dfrac{2}{\phi} - 1$. Similarly, we start with the expression $\dfrac{1}{\phi^2} = \dfrac{1}{\phi^3} + \dfrac{1}{\phi^4}$ and substitute $\dfrac{2}{\phi} - 1$ for $\dfrac{1}{\phi^3}$, and $1 - \dfrac{1}{\phi}$ for $\dfrac{1}{\phi^2}$ to obtain

$$\frac{1}{\phi^4} = \frac{1}{\phi^2} - \frac{1}{\phi^3} = \left(1 - \frac{1}{\phi}\right) - \left(\frac{2}{\phi} - 1\right) = 2 - \frac{3}{\phi}.$$

We can continue in a similar fashion to prove the other expressions shown above and generate the Fibonacci numbers in this way.

There are more interesting properties of the Golden Ratio ϕ. For example, we can verify the following identities involving ϕ:

$$\frac{1}{\phi} + \frac{1}{\phi^2} = 1$$

$$\phi + \frac{1}{\phi^2} = 2$$

$$\phi^2 + \frac{1}{\phi^2} = 3$$

$$\left(\phi+\frac{1}{\phi^2}\right)^2 = 4$$

$$\left(\phi+\frac{1}{\phi}\right)^2 = 5$$

The above five identities are verified using MATLAB as follows:

```
>> phi = (1+sqrt(5))/2

phi =

    1.6180

>> 1/phi + 1/phi^2

ans =

     1

>> phi + 1/phi^2

ans =

     2

>> phi^2 + 1/phi^2

ans =

     3

>> (phi + 1/phi^2)^2
```

```
ans =

     4

>> (phi + 1/phi)^2

ans =

    5.0000
```

Another interesting property is that $\phi + \dfrac{1}{\phi} = \sqrt{5}$. This property is verified using MATLAB as follows:

```
>> phi = (1+sqrt(5))/2

phi =

    1.6180

>> phi + 1/phi

ans =

    2.2361

>> sqrt(5)

ans =

    2.2361
```

Finally, we list the following very interesting property of ϕ :

$$\phi = \frac{1}{\phi} + \frac{1}{\phi^2} + \frac{1}{\phi^3} + \frac{1}{\phi^4} + \frac{1}{\phi^5} + \frac{1}{\phi^6} + \ldots\ldots$$

We verify the above property using MATLAB as follows:

```
>> phi = (1+sqrt(5))/2

phi =

    1.6180

>> 1/phi + 1/phi^2 + 1/phi^3 + 1/phi^4 +
1/phi^5 + 1/phi^6

ans =

    1.5279
```

Obviously, we did not obtain the exact result above but we obtained an approximate value. This is because we used only the first six terms in the series. We would have obtained more precision and a more exact value for ϕ if we used more terms in the series.

The Golden Ratio can also be written as a continued fraction as follows

$$\phi = 1 + \cfrac{1}{1 + \cfrac{1}{1 + \cfrac{1}{1 + \ldots}}}$$

The above formula can be show to be valid easily as follows. It is very clear that the main denominator of the fraction on the right-hand-side is equal to ϕ also. Therefore, the above equation for the continue fraction is in the form $\phi = 1 + \dfrac{1}{\phi}$. This equation when simplified reduces to the familiar quadratic equation $\phi^2 = \phi + 1$ that is associated with the Golden Ratio.

We devote the remaining part of this chapter to deriving a compact equation for the computation of Fibonacci numbers using the value of the Golden Ratio. Let us start with the quadratic equation for the Golden Ratio that was derived in the previous chapter:

$$x^2 = x + 1$$

It was illustrated in the previous chapter that the two roots of the above equation are ϕ and $1 - \phi$. Substituting each value in the above quadratic equation separately, we obtain the following two equations:

$$\phi^2 = \phi + 1$$

$$(1 - \phi)^2 = (1 - \phi) + 1$$

Next, we multiply both sides of the first equation above by ϕ^n, and multiply both sides of the second equation above by $(1 - \phi)^n$ to obtain the following two equations:

$$\phi^{n+2} = \phi^{n+1} + \phi^n$$

$$(1 - \phi)^{n+2} = (1 - \phi)^{n+1} + (1 - \phi)^n$$

where $n = 0, 1, 2, 3, \dots$. Next, we subtract the second equation from the first equation above to obtain:

$$\phi^{n+2} - (1-\phi)^{n+2} = \phi^{n+1} - (1-\phi)^{n+1} + \phi^{n} - (1-\phi)^{n}$$

Next, we divide both sides of the above equation by $\phi - (1-\phi)$[26] to obtain the following recursive relation:

$$\frac{\phi^{n+2} - (1-\phi)^{n+2}}{\phi - (1-\phi)} = \frac{\phi^{n+1} - (1-\phi)^{n+1}}{\phi - (1-\phi)} + \frac{\phi^{n} - (1-\phi)^{n}}{\phi - (1-\phi)}$$

The above equation is a recursive relation which can be written in the general mathematical notation:

$$F(n+2) = F(n+1) + F(n)$$

It is seen that the above recursive relation is precisely the equation that governs the sequence of Fibonacci numbers (see the chapter on Fibonacci numbers). Therefore, by comparing the above two equations above, we can conclude that the general formula for $F(n)$ can be written as follows:

$$F(n) = \frac{\phi^{n} - (1-\phi)^{n}}{\phi - (1-\phi)} \quad , \quad n = 0,1,2,3,....$$

The above general equation can be then simplified as follows:

$$F(n) = \frac{\phi^{n} - (1-\phi)^{n}}{2\phi - 1} \quad , \quad n = 0,1,2,3,...$$

Thus, we conclude that the above equation can be used to computer Fibonacci numbers using the value of the Golden

---

[26] This expression is equal to $2\phi - 1$ but will keep it in this unsimplified form in order to derive the sought relationship between Fibonacci numbers and the Golden Ratio.

Ratio, i.e. the n-th term $F(n)$ of the Fibonacci sequence can be computed using the above general relation using the value of ϕ only. We can simplify the above general equation further by substituting the value $\dfrac{1+\sqrt{5}}{2}$ for ϕ in order to obtain:

$$F(n) = \frac{\left(1+\sqrt{5}\right)^n - \left(1-\sqrt{5}\right)^n}{2^n \sqrt{5}} \quad , \quad n = 0,1,2,3,\ldots.$$

Thus, the above formula is the general expression for computing any Fibonacci number without generating the whole sequence. In order to use the above equation, we only need to insert the value of n, i.e. we only need to know which Fibonacci number we need to compute.

Finally, we will write a short MATLAB function which uses the above formula to compute any Fibonacci number. The MATLAB code for the function fib(n) is shown below:

```
function f=fib(n)
%  This  function  generates  the  nth
Fibonacci %  number
f  =  ((1+sqrt(5))^n  -(1-sqrt(5))^n)
/(sqrt(5)*2^n);
```

Next, let us execute the above MATLAB function several times in order to computer a few Fibonacci numbers. Here is the MATLAB output:

```
>> fib(5)

ans =
```

```
     5

>> fib(7)

ans =

    13

>> fib(10)

ans =

   55.0000

>> fib(15)

ans =

  610.0000
```

Thus, the MATLAB function `fib(n)` works perfectly. Just insert any value for `n` and execute the function `fib(n)` to get the desired nth term of the Fibonacci sequence without generating the sequence itself. We have been able to do this because we used the Golden Ratio.

It should be mentioned finally that there is an approximation for the above formula for calculating Fibonacci numbers.

Instead of using the above long formula, we can use the following shorter approximate formula:

$$F(n) \approx \frac{\phi^n}{\sqrt{5}}$$

Substituting the value of ϕ which is $\dfrac{1+\sqrt{5}}{2}$ in the above formula, we obtain the following explicit expression for the approximate formula:

$$F(n) \approx \frac{\left(1+\sqrt{5}\right)^n}{2^n \sqrt{5}}$$

It should be noted that the above approximate formula gives more accurate results for large value of n, i.e. for large Fibonacci numbers. The approximate short formula can be derived from the exact long formula by taking the limit[27] of the exact formula as n approaches infinity. We now write a new MATLAB function called `approximate_fib(n)` to calculate Fibonacci numbers using the approximate formula. Here is the MATLAB code for this function:

```
function f=approximate_fib(n)
%  This    function   generates   the    nth
Fibonacci %  number using the approximate
formula
f = ((1+sqrt(5))^n) /(sqrt(5)*2^n);
```

We execute the MATLAB function `approximate_fib(n)` for the first five n values 1, 2, 3, 4, 5 to obtain the following approximate values for Fibonacci numbers:

```
>> approximate_fib(1)

ans =
```

---

[27] Note that $\displaystyle\lim_{n\to\infty}\frac{\phi^n-\left(1-\phi\right)^n}{2\phi-1}=\frac{\phi^n}{\sqrt{5}}$. You need to revise your calculus information in order to derive this limit.

```
     0.7236

>> approximate_fib(2)

ans =

     1.1708

>> approximate_fib(3)

ans =

     1.8944

>> approximate_fib(4)

ans =

     3.0652

>> approximate_fib(5)

ans =

     4.9597
```

Next, we execute the above MATLAB function several times and compare its results with those of the exact function `fib(n)` as follows (we see that we obtain exact results for higher values of n):

```
>> fib(10)

ans =

    55.0000
```

```
>> approximate_fib(10)

ans =

    55.0036

>> fib(30)

ans =

  8.3204e+005

>> approximate_fib(30)

ans =

  8.3204e+005
```

Thus, for the tenth Fibonacci number, we obtain almost an exact result. We obtain also an almost exact result for the 30th Fibonacci number as shown above. The larger the value of n, the more accurate is the Fibonacci number which is obtained.

In the next chapter, we study another sequence of interesting numbers that is closely related to the Fibonacci sequence.

References

Books on Fibonacci Numbers and Golden Ratio

1. Posamentier, A. S., *The Fabulous Fibonacci Numbers*, Prometheus Books, 2007.

2. Vajda, S., *Fibonacci and Lucas Numbers, and the Golden Section*, Dover Publications, 2007.

3. Vorobiev, N. N. and Martin, M., *Fibonacci Numbers*, Birkhauser, 2003.

4. Dunlap, R. A., *The Golden Ratio and Fibonacci Numbers*, World Scientific Publishing, 1997.

5. Olsen, S., *The Golden Section: Nature's Greatest Secret*, Walker and Company, 2006.

6. Livio, M., *The Golden Ratio: The Story of PHI, the World's Most Astonishing Number*, Broadway, 2003.

7. Koshy, T., *Fibonacci and Lucas Numbers with Applications*, Wiley-Interscience, 2001.

8. Dobson, E. D., *Understanding Fibonacci Numbers*, Traders Press, 1994.

9. Hoggatt, V. E. and Bicknell, M., *Primer on Fibonacci Numbers*, Fibonacci Association, 1972.

Web Links for Fibonacci Numbers and Golden Ratio

10. Fibonacci Numbers at Wolfram MathWorld
http://mathworld.wolfram.com/FibonacciNumber.html

11. Fibonacci Numbers at Wikipedia
http://en.wikipedia.org/wiki/Fibonacci_number

12. Fibonacci Numbers and the Golden Section

http://www.maths.surrey.ac.uk/hosted-sites/R.Knott/Fibonacci/fib.html

13. The Mathematical Magic of Fibonacci Numbers
http://www.maths.surrey.ac.uk/hosted-sites/R.Knott/Fibonacci/fibmaths.html

14. The Fibonacci Sequence – Math is Fun
http://www.mathsisfun.com/numbers/fibonacci-sequence.html

15. Generalizations of Fibonacci Numbers from Wikipedia
http://en.wikipedia.org/wiki/Tribonacci_number

16. Random Fibonacci Sequences from Wikipedia
http://en.wikipedia.org/wiki/Viswanath%27s_constant

17. The Golden Mean and the Physics of Aesthetics by Subhash Kak
http://arxiv.org/abs/physics/0411195

18. Stepping Beyond Fibonacci Numbers – from Ivars Peterson's MathTrek
http://www.maa.org/mathland/mathtrek_09_30_02.html

19. Viswanath's Constant from PlanetMath
http://planetmath.org/encyclopedia/ViswanathsConstant.html

20. Divakar Viswanath, "Random Fibonacci Sequences and the Number 1.13198824…", Mathematics of Computation, Vol. 69, 2000, pp. 1131-1155.

7. Regression Analysis with MATLAB

Regression analysis is a statistical method which helps to determine an approximate form of the relationship between variables. In this chapter[28] we consider the relationship between two variables using simple linear regression[29].

Two variables are related linearly if their relationship can be expressed by the equation of a straight line. This is the essence of simple linear regression. The resulting straight line is considered to be a best-fit for the data under study. We assume the equation of the straight line to be written in the following form:

$$y = a + bx$$

where a and b are constant coefficients to be determined for a specific set of data (the coefficient b represents the slope or gradient of the resulting straight line while the coefficient a represents the y-intercept, i.e. the point where the resulting straight line intersects the y-axis). We will proceed with the detailed derivation of the equations of simple linear regression.

For a set of data x_i and y_i (where $i = 1, 2, 3, \ldots n$), the above equation becomes for each point (x_i, y_i):

$$y_i = a + bx_i \qquad\qquad \text{(A)}$$

---

[28] This section appeared originally as a chapter in the book "MATLAB for Beginners: A Gentle Approach – Second Edition."

[29] There are some specialized MATLAB commands like the command `polyfit` that can be used to find the best fitting polynomial or straight line for a given set of data. But such specialized and advanced commands are not used in this book. Instead use is made of the various MATLAB commands found in the basic package to solve these types of problems. In addition, using simple basic commands from scratch enhances the learning process considerably.

It should be noted that the above equation is repeated n times, where n is the number of data points. Let us write the above equations for the first few data points as $i = 1, 2, 3, \ldots\ldots n$:

$$y_1 = a + bx_1$$

$$y_2 = a + bx_2$$

$$y_3 = a + bx_3$$

$$\ldots\ldots\ldots\ldots\ldots$$

$$\ldots\ldots\ldots\ldots\ldots$$

$$\ldots\ldots\ldots\ldots\ldots$$

$$y_n = a + bx_n$$

Let us now add the above list of equations to obtain the following expression:

$$y_1 + y_2 + y_3 + \ldots\ldots + y_n = a + bx_1 + a + bx_2 + a + bx_3 + \ldots\ldots + a + bx_n$$

The above equation can be written in the following simplified form:

$$y_1 + y_2 + y_3 + \ldots\ldots + y_n = \left(a + a + a + \ldots\ldots + a\right) + \left(bx_1 + bx_2 + bx_3 + \ldots\ldots + bx_n\right)$$

where the sum of the a's in the first parenthesis is n times. The above equation can be written using the summation sign $\sum$ as follows:

$$\sum_{i=1}^{n} y_i = na + b\sum_{i=1}^{n} x_i \tag{B}$$

where the following summation notation is used in equation (B):

$$\sum_{i=1}^{n} y_i = y_1 + y_2 + y_3 + \ldots\ldots + y_n$$

$$na = a + a + a + \ldots + a \qquad (summed\ \ n\ \ times)$$

$$b\sum_{i=1}^{n} x_i = b\left(x_1 + x_2 + x_3 + \ldots\ldots + x_n\right) = bx_1 + bx_2 + bx_3 + \ldots + bx_n$$

The next step would be to multiply equation (A) by x_i to obtain the following equation:

$$x_i y_i = ax_i + bx_i x_i \tag{C}$$

Writing the above equation for each value $i = 1, 2, 3, \ldots\ldots n$, we obtain:

$$x_1 y_1 = ax_1 + bx_1^2$$

$$x_2 y_2 = ax_2 + bx_2^2$$

$$x_3 y_3 = ax_3 + bx_3^2$$

$$\ldots\ldots\ldots\ldots\ldots$$

$$\ldots\ldots\ldots\ldots\ldots$$

$$\ldots\ldots\ldots\ldots\ldots$$

$$x_n y_n = ax_n + bx_n^2$$

Adding the above list of equations results in the following expression:

$$x_1 y_1 + x_2 y_2 + x_3 y_3 + \ldots + x_n y_n = ax_1 + bx_1^2 + ax_2 + bx_2^2 + ax_3 + bx_3^2$$
$$+ \ldots + ax_n + bx_n^2$$

The above equation can be simplified as follows:

$$x_1 y_1 + x_2 y_2 + x_3 y_3 + \ldots + x_n y_n = ax_1 + ax_2 + ax_3 + \ldots + ax_n$$
$$+ bx_1^2 + bx_2^2 + bx_3^2 + \ldots + bx_n^2$$

The above equation can be written using the summation sign $\sum$ as follows:

$$\sum_{i=1}^{n} x_i y_i = a \sum_{i=1}^{n} x_i + b \sum_{i=1}^{n} x_i^2 \qquad (D)$$

where the following summation notation is used in equation (D):

$$\sum_{i=1}^{n} x_i y_i = x_1 y_1 + x_2 y_2 + x_3 y_3 + \ldots + x_n y_n$$

$$a \sum_{i=1}^{n} x_i = a\left(x_1 + x_2 + x_3 + \ldots + x_n\right) = ax_1 + ax_2 + ax_3 + \ldots + ax_n$$

$$b\sum_{i=1}^{n} x_i^2 = b\left(x_1^2 + x_2^2 + x_3^2 + \ldots\ldots + x_n^2\right) = bx_1^2 + bx_2^2 + bx_3^2 + \ldots\ldots + bx_n^2$$

Next we solve equations (B) and (D) simultaneously. The two equations are summarized below followed by their solution which was obtained using the method of elimination manually or using MATLAB with the command `solve` as illustrated in Chapter 10 on Solving Equations:

$$\sum_{i=1}^{n} y_i = na + b\sum_{i=1}^{n} x_i$$

$$\sum_{i=1}^{n} x_i y_i = a\sum_{i=1}^{n} x_i + b\sum_{i=1}^{n} x_i^2$$

The solution of the above two linear equations for the constants a and b is obtained as follows:

$$b = \frac{n\sum_{i=1}^{n} x_i y_i - \sum_{i=1}^{n} x_i \sum_{i=1}^{n} y_i}{n\sum_{i=1}^{n} x_i^2 - \left(\sum_{i=1}^{n} x_i\right)^2} \qquad\qquad \text{(I)}$$

$$a = \frac{\sum_{i=1}^{n} x_i^2 \sum_{i=1}^{n} y_i - \sum_{i=1}^{n} x_i \sum_{i=1}^{n} x_i y_i}{n\sum_{i=1}^{n} x_i^2 - \left(\sum_{i=1}^{n} x_i\right)^2} \qquad\qquad \text{(II)}$$

The procedure for curve fitting a straight line through the method of simple linear regression based on the above two equations is summarized in the following five steps:

1. Calculate the four sums $\sum_{i=1}^{n} x_i$, $\sum_{i=1}^{n} y_i$, $\sum_{i=1}^{n} x_i^2$, and $\sum_{i-1}^{n} x_i y_i$ using MATLAB (if calculated manually, it is preferred to do it in a table). Table 1 shows each sum above along with the needed MATLAB command to compute it.

Table 1: MALTAB Command Needed to Compute Each Sum

Sum	MATLAB Command
x_i	X
y_i	Y
$\sum_{i=1}^{n} x_i$	sum(x)
$\sum_{i=1}^{n} y_i$	sum(y)
$\sum_{i=1}^{n} x_i^2$	mxx = x.*x sum(mxx)
$\sum_{i=1}^{n} x_i y_i$	mxy = x.*y sum(mxy)
$\left(\sum_{i=1}^{n} x_i\right)^2$	(sum(x))^2

2. Substitute the above four sums into equations (I) and (II) to obtain the two constants a and b.

3. Using the two values of the constants a and b obtained above, write the equation of the best fitting line as follows:

$$y = a + bx$$

4. To check the results or to view the results graphically, plot the data using the MATLAB command `scatter`, then plot the best fitting line on the same graph using the MATLAB command `plot`. These two commands will be illustrated in the numerical example that follows.

5. This last step is optional. We can calculate various types of errors between the data and the best fitting line, including the maximum error. However, this step will not be illustrated in the numerical example that follows.

In order to illustrate simple linear regression in this chapter, we employ what is known as the *Least Squares* procedure. This procedure is used to obtain a best-fit curve or a best-fit line that represents the data based on simple linear regression. This is shown below using an example.

Let us now solve a numerical example using the above five steps. The following data (see Table 2) show the monthly salaries (x, in dollars) and the amount of monthly rent payments (y, in dollars) for a random sample of 16 employees at a certain company. Let us obtain the equation of the best fitting line using simple linear regression.

Table 2: Data for the Numerical Example

x	600	700	650	680	750	800	900	850
y	150	180	200	230	250	280	300	350

Table 2: Data for the Numerical Example (Continued)

x	95 0	100 0	102 0	105 0	110 0	120 0	125 0	130 0

y	400	430	450	480	500	510	550	560

Let us now proceed to start the solution of the numerical example given above using the five steps outlined above for simple linear regression.

Step 1: Calculate the four sums $\sum_{i=1}^{n} x_i$, $\sum_{i=1}^{n} y_i$, $\sum_{i=1}^{n} x_i^2$, and $\sum_{i=1}^{n} x_i y_i$ using the MATLAB command sum[30] as follows:

1.1: Let us first input the data as follows:

```
>> x = [600 700 650 680 750 800 900 850 950
1000 1020 1050 1100 1200 1250 1300]

x =

   Columns 1 through 6

        600         700         650         680
750         800

   Columns 7 through 12

        900         850         950        1000
1020         1050

   Columns 13 through 16

       1100        1200        1250        1300

>> y = [150 180 200 230 250 280 300 350 400
430 450 480 500 510 550 560]

y =
```

---

[30] For more details about the MATLAB command sum, see Chapter 6 on Vectors.

```
Columns 1 through 12

   150    180    200    230    250    280    300
350  400  430  450  480
```

```
Columns 13 through 16

  500   510   550   560
```

1.2: Let us calculate the sum $\sum_{i=1}^{n} x_i$:

```
>> sumx = sum(x)

sumx =

     14800
```

1.3: Let us calculate the sum $\sum_{i=1}^{n} y_i$:

```
>> sumy = sum(y)

sumy =

     5820
```

1.4: Let us calculate the sum $\sum_{i=1}^{n} x_i^2$:

```
>> mxx = x.*x

mxx =

   Columns 1 through 5

      360000     490000     422500     462400
562500

   Columns 6 through 10

      640000     810000     722500     902500
1000000
```

```
    Columns 11 through 15

        1040400      1102500      1210000      1440000
1562500

    Column 16

        1690000

>> sumxx = sum(mxx)

sumxx =

    14417800
```

1.5: Let us calculate the sum $\displaystyle\sum_{i=1}^{n} x_i y_i$:

```
>> mxy = x.*y

mxy =

    Columns 1 through 5

          90000        126000        130000        156400
187500

    Columns 6 through 10

         224000        270000        297500        380000
430000

    Columns 11 through 15

         459000        504000        550000        612000
687500

    Column 16

         728000

>> sumxy = sum(mxy)

sumxy =
```

```
    5831900
```

1.6: Let us calculate the square $\left(\sum\limits_{i=1}^{n} x_i \right)^2$:

```
>> sumx2 = sumx*sumx

sumx2 =

    219040000
```

1.7: Let us calculate the number of data points using the `size` command:

```
>> n = size(x)

n =

     1     16

>> n1 = n(2)

n1 =

    16
```

<u>Step 2</u>: Substitute the values of the above four sums into equations (I) and (II), evaluate the two expressions in these two equations, and obtain the values of the constants a and b. This is performed in MATLAB as follows:

```
>> a = (sumxx*sumy - sumx*sumxy)/(n1*sumxx - sumx2)

a =

   -206.1456

>> b = (n1*sumxy - sumx*sumy)/(n1*sumxx - sumx2)

b =
```

```
0.6161
```

<u>Step 3</u>: Using the values of the constants a and b obtained above, we can finally write the equation of the best fitting line as follows:

$$y = -206.1456 + 0.6161\,x$$

<u>Step 4</u>: In order to compare the best fitting line with the data given, let us know plot the data along with the best fitting line on the same graph. This is performed using the two MATLAB commands `scatter` and `plot` as follows:

4.1: Let us calculate the y values using the best fit equation above as follows:

```
>> y1 = a + b*x

y1 =

  Columns 1 through 7

   163.5164    225.1268    194.3216    212.8047
255.9319   286.7371   348.3474

  Columns 8 through 14

   317.5423    379.1526    409.9577    422.2798
440.7629   471.5681   533.1784

  Columns 15 through 16

  563.9836   594.7887
```

4.2: Let us plot the best fitting straight line:

```
>> plot(x,y1)
>> hold on
```

4.3: Let us plot the given data points:

```
>> scatter(x,y)
```

4.4: Let us label the x-axis and the y-axis:

```
>> xlabel('x')
>> ylabel('y')
```

The graph generated by MATLAB is shown in Figure 1. It is clear from the figure that the generated best fitting line truly represents the data points given in the example.

Step 5: This error calculation step is optional and is not demonstrated here.

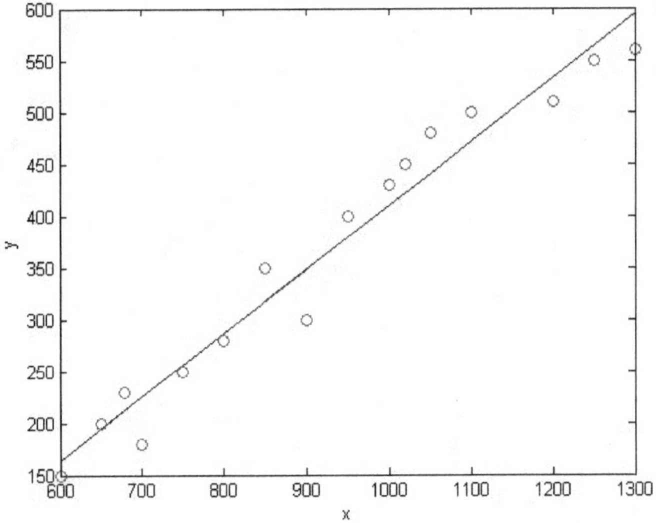

Figure 1: Best Fitting Line Along with the Given Data Points

for the Example

Exercises

Solve the following three exercises on simple regression analysis. The first two should be done with simple linear regression using the five steps presented in this chapter. However, the third exercise is different. The third exercise is on the topic of simple nonlinear regression (specifically simple quadratic or simple parabolic regression) that is not illustrated in this chapter. However, enough information, hints, and steps are given in the exercise to make its solution attainable by the student.

1. A company makes a study of its assets (x, in thousands of dollars) and its expenditure (y, in thousands of dollars) for a specific department. The following data are given (see Table 3):

Table 3: Data for Exercise 1

x	50	52	56	61	66	70	80	84	88	92
y	200	220	250	310	360	400	420	440	500	550

Find the equation of the best fitting for the above data. Plot the data along with the best fitting line on the same graph.

2. The following data (see Table 4) shows the expenditure x and the net income y, both in dollars, for a random sample of 12 firms.

Table 4: Data for Exercise 2

x	10	20	25	31	36	44	55	60	62	65	70	75
y	15	25	30	35	40	50	60	65	70	75	80	85

Find the equation of the best fitting for the above data. Plot the data along with the best fitting line on the same graph.

3. This exercise is optional for the beginning student and involves the equations of simple nonlinear regression, specifically what is called simple quadratic regression (also called simple parabolic regression). We will not solve a numerical problem in this exercise, but only want to write the equations for this special case of simple nonlinear regression.

For a given set of data (x_i, y_i), assume the equation of the best fitting parabola to be quadratic in the general form:

$$y = a + bx + cx^2$$

Derive the three algebraic equations for the constant coefficients a, b, and c using the method of simple nonlinear regression.

Solutions to the Exercises

1. <u>Step One:</u> Enter the given data points and calculate the different required sums:

```
>> x = [50 52 56 61 66 70 80 84 88 92]

x =

    50    52    56    61    66    70    80    84
88    92

>> y = [200 220 250 310 360 400 420 440 500 550]

y =
```

```
    200    220    250    310    360    400    420    440
500    550

>> sumx = sum(x)

sumx =

   699

>> sumy = sum(y)

sumy =

        3650

>> mxx = x.*x

mxx =

  Columns 1 through 6

        2500        2704        3136        3721
4356        4900

  Columns 7 through 10

        6400        7056        7744        8464

>> sumxx = sum(mxx)

sumxx =

        50981

>> mxy = x.*y

mxy =

  Columns 1 through 5

        10000       11440       14000       18910
23760

  Columns 6 through 10
```

```
        28000           33600           36960           44000
50600

>> sumxy = sum(mxy)

sumxy =

      271270

>> sumx2 = sumx*sumx

sumx2 =

      488601

>> n = size(x)
n =

      1      10

>> n1 = n(2)

n1 =

     10
```

<u>Step Two:</u> Calculate the two constants a and b using the two formulas given:

```
>> a = (sumxx*sumy - sumx*sumxy)/(n1*sumxx - sumx2)

a =

 -166.7726

>> b = (n1*sumxy - sumx*sumy)/(n1*sumxx - sumx2)

b =

    7.6076
```

<u>Step Three:</u> Write the equation of the best fitting line as follows:

$$y = -166.7726 + 7.6076\,x$$

<u>Step Four:</u> Plot the best fitting line along with the give data points on the same graph for comparison:

```
>> y1 = a + b*x

y1 =

  Columns 1 through 7

   213.6084      228.8236      259.2541      297.2922
335.3303   365.7608   441.8370

  Columns 8 through 10

   472.2674   502.6979   533.1284

>> plot(x,y1)
>> hold on
>> scatter(x,y)
>> xlabel('x')
>> ylabel('y')
```

The resulting graph is shown in the figure below.

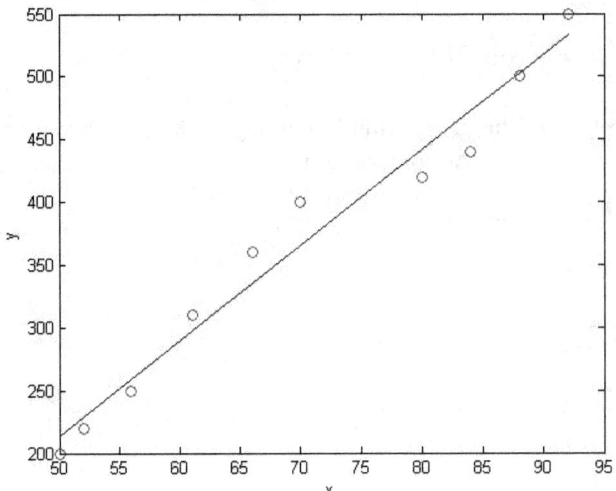

2. <u>Step One:</u> Enter the given data points and calculate the different sums needed:

```
>> x = [10 20 25 31 36 44 55 60 62 65 70 75]

x =

    10    20    25    31    36    44    55    60
62    65    70    75

>> y = [15 25 30 35 40 50 60 65 70 75 80 85]

y =

    15    25    30    35    40    50    60    65
70    75    80    85

>> sumx = sum(x)

sumx =

   553

>> sumy = sum(y)
```

```
sumy =

   630

>> mxx = x.*x

mxx =

  Columns 1 through 6

         100         400         625         961
1296        1936

  Columns 7 through 12

        3025        3600        3844        4225
4900        5625

>> sumxx = sum(mxx)

sumxx =

       30537

>> mxy = x.*y

mxy =

  Columns 1 through 5

         150         500         750        1085
1440

  Columns 6 through 10

        2200        3300        3900        4340
4875

  Columns 11 through 12

        5600        6375

>> sumxy = sum(mxy)
```

```
sumxy =

        34515

>> sumx2 = sumx*sumx

sumx2 =

        305809

>> n = size(x)

n =

        1       12

>> n1 = n(2)

n1 =

        12
```

Step Two: Calculate the two constants a and b as follows:

```
>> a = (sumxx*sumy - sumx*sumxy)/(n1*sumxx -
sumx2)

a =

    2.4988

>> b = (n1*sumxy - sumx*sumy)/(n1*sumxx - sumx2)

b =

    1.0850
```

Step 3: Write the equation of the best fitting line as follows:

$$y = 2.4988 + 1.0850\,x$$

Step 4: Plot the best fitting line along with the given date points on the same graph for comparison:

```
>> y1 = a + b*x

y1 =

  Columns 1 through 7

   13.3490    24.1991    29.6242    36.1343    41.5594
50.2395    62.1747

  Columns 8 through 12

   67.5998    69.7699    73.0249    78.4500    83.8751

>> plot(x,y1)
>> hold on
>> scatter(x,y)
>> xlabel('x')
>> ylabel('y')
```

The resulting graph is shown in the figure below.

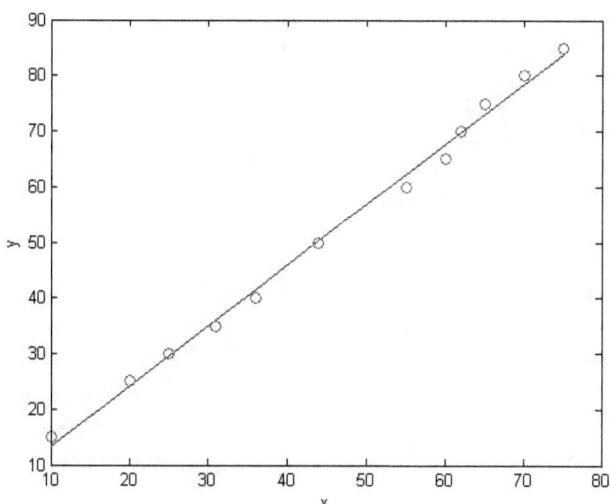

3..   Let us start with the quadratic equation given in the Exercise
       as follows:

$$y = a + bx + cx^2$$

where a, b, and c are constant coefficients to be determined for a specific set of data. We will proceed with the detailed derivation of the equations of simple nonlinear regression.

For a set of data x_i and y_i (where $i = 1,2,3,......n$), the above equation becomes for each point (x_i, y_i):

$$y_i = a + bx_i + cx_i^2 \qquad\qquad \text{(AA)}$$

It should be noted that the above equations is repeated n times, where n is the number of data points. Let us write the above equations for the first few data points as $i = 1,2,3,......n$:

$$y_1 = a + bx_1 + cx_1^2$$

$$y_2 = a + bx_2 + cx_2^2$$

$$y_3 = a + bx_3 + cx_3^2$$

$$\ldots\ldots\ldots\ldots$$

$$\ldots\ldots\ldots\ldots$$

$$\ldots\ldots\ldots\ldots$$

$$y_n = a + bx_n + cx_n^2$$

Let us now add the above list of equations to obtain the following expression:

$$y_1 + y_2 + y_3 + ... + y_n = a + bx_1 + cx_1^2 + a + bx_2 + cx_2^2$$
$$+ a + bx_3 + cx_3^2 + ... + a + bx_n + cx_n^2$$

The above equation can be written in the following simplified form:

$$y_1 + y_2 + y_3 + ... + y_n = (a + a + a + ... + a)$$
$$+ (bx_1 + bx_2 + bx_3 + ... + bx_n) + (cx_1^2 + cx_2^2 + cx_3^2 + ... + cx_n^2)$$

where the sum of the a's in the first parenthesis is n times. The above equation can be written using the summation sign $\sum$ as follows:

$$\sum_{i=1}^{n} y_i = na + b\sum_{i=1}^{n} x_i + c\sum_{i=1}^{n} x_i^2 \qquad \text{(BB)}$$

where the following summation notation is used in equation (BB):

$$\sum_{i=1}^{n} y_i = y_1 + y_2 + y_3 + ... + y_n$$

$$na = a + a + a + ... + a \qquad (summed\ n\ times)$$

$$b\sum_{i=1}^{n} x_i = b(x_1 + x_2 + x_3 + ... + x_n) = bx_1 + bx_2 + bx_3 + ... + bx_n$$

$$c\sum_{i=1}^{n} x_i^2 = c(x_1^2 + x_2^2 + x_3^2 + ... + x_n^2) = cx_1^2 + cx_2^2 + cx_3^2 + ... + cx_n^2$$

The next step would be to multiply equation (AA) by x_i to obtain the following equation:

$$x_i y_i = ax_i + bx_i x_i + cx_i^2 x_i = ax_i + bx_i^2 + cx_i^3 \qquad \text{(CC)}$$

Writing the above equation out for each value $i = 1, 2, 3, \ldots\ldots n$, we obtain:

$$x_1 y_1 = ax_1 + bx_1^2 + cx_1^3$$

$$x_2 y_2 = ax_2 + bx_2^2 + cx_2^3$$

$$x_3 y_3 = ax_3 + bx_3^2 + cx_3^3$$

$$\ldots\ldots\ldots\ldots\ldots\ldots$$

$$\ldots\ldots\ldots\ldots\ldots\ldots$$

$$\ldots\ldots\ldots\ldots\ldots\ldots$$

$$x_n y_n = ax_n + bx_n^2 + cx_n^3$$

Adding the above list of equations results in the following expression:

$$x_1 y_1 + x_2 y_2 + x_3 y_3 + \ldots\ldots + x_n y_n = ax_1 + bx_1^2 + cx_1^3 + ax_2 + bx_2^2 + cx_2^3$$
$$+ ax_3 + bx_3^2 + cx_3^3 + \ldots\ldots\ldots + ax_n + bx_n^2 + cx_n^3$$

The above equation can be simplified as follows:

$$x_1 y_1 + x_2 y_2 + x_3 y_3 + \ldots\ldots + x_n y_n = ax_1 + ax_2 + ax_3 + \ldots\ldots + ax_n$$
$$+ bx_1^2 + bx_2^2 + bx_3^2 + \ldots\ldots + bx_n^2 + cx_1^3 + cx_2^3 + cx_3^3 + \ldots\ldots + cx_n^3$$

The above equation can be written using the summation sign $\sum$ as follows:

$$\sum_{i=1}^{n} x_i y_i = a\sum_{i=1}^{n} x_i + b\sum_{i=1}^{n} x_i^2 + c\sum_{i=1}^{n} x_i^3 \qquad \text{(DD)}$$

where the following summation notation is used in equation (DD):

$$\sum_{i=1}^{n} x_i y_i = x_1 y_1 + x_2 y_2 + x_3 y_3 + \ldots + x_n y_n$$

$$a\sum_{i=1}^{n} x_i = a\left(x_1 + x_2 + x_3 + \ldots + x_n\right) = ax_1 + ax_2 + ax_3 + \ldots + ax_n$$

$$b\sum_{i=1}^{n} x_i^2 = b\left(x_1^2 + x_2^2 + x_3^2 + \ldots + x_n^2\right) = bx_1^2 + bx_2^2 + bx_3^2 + \ldots + bx_n^2$$

$$c\sum_{i=1}^{n} x_i^3 = c\left(x_1^3 + x_2^3 + x_3^3 + \ldots + x_n^3\right) = cx_1^3 + cx_2^3 + cx_3^3 + \ldots + cx_n^3$$

The next step would be to multiply equation (AA) by x_i^2 to obtain the following equation:

$$x_i^2 y_i = ax_i^2 + bx_i x_i^2 + cx_i^2 x_i^2 = ax_i^2 + bx_i^3 + cx_i^4 \quad \text{(EE)}$$

Writing the above equation out for each value $i = 1, 2, 3, \ldots n$, we obtain:

$$x_1^2 y_1 = ax_1^2 + bx_1^3 + cx_1^4$$

$$x_2^2 y_2 = ax_2^2 + bx_2^3 + cx_2^4$$

$$x_3^2 y_3 = ax_3^2 + bx_3^3 + cx_3^4$$

$$.................$$

$$.................$$

$$.................$$

$$x_n^2 y_n = ax_n^2 + bx_n^3 + cx_n^4$$

Adding the above list of equations results in the following expression:

$$x_1^2 y_1 + x_2^2 y_2 + x_3^2 y_3 + ... + x_n^2 y_n = ax_1^2 + bx_1^3 + cx_1^4 + ax_2^2 + bx_2^3 + cx_2^4$$
$$+ ax_3^2 + bx_3^3 + cx_3^4 + ... + ax_n^2 + bx_n^3 + cx_n^4$$

The above equation can be simplified as follows:

$$x_1^2 y_1 + x_2^2 y_2 + x_3^2 y_3 + ... + x_n^2 y_n = ax_1^2 + ax_2^2 + ax_3^2 + ... + ax_n^2$$
$$+ bx_1^3 + bx_2^3 + bx_3^3 + ... + bx_n^3 + cx_1^4 + cx_2^4 + cx_3^4 + ... + cx_n^4$$

The above equation can be written using the summation sign $\sum$ as follows:

$$\sum_{i=1}^{n} x_i^2 y_i = a \sum_{i=1}^{n} x_i^2 + b \sum_{i=1}^{n} x_i^3 + c \sum_{i=1}^{n} x_i^4 \qquad \text{(FF)}$$

where the following summation notation is used in equation (FF):

$$\sum_{i=1}^{n} x_i^2 y_i = x_1^2 y_1 + x_2^2 y_2 + x_3^2 y_3 + \ldots\ldots + x_n^2 y_n$$

$$a\sum_{i=1}^{n} x_i^2 = a\left(x_1^2 + x_2^2 + x_3^2 + \ldots\ldots + x_n^2\right) = ax_1^2 + ax_2^2 + ax_3^2 + \ldots\ldots + ax_n^2$$

$$b\sum_{i=1}^{n} x_i^3 = b\left(x_1^3 + x_2^3 + x_3^3 + \ldots\ldots + x_n^3\right) = bx_1^3 + bx_2^3 + bx_3^3 + \ldots\ldots + bx_n^3$$

$$c\sum_{i=1}^{n} x_i^4 = c\left(x_1^4 + x_2^4 + x_3^4 + \ldots\ldots + x_n^4\right) = cx_1^4 + cx_2^4 + cx_3^4 + \ldots\ldots + cx_n^4$$

Next we solve equations (BB), (DD), and (FF) simultaneously. The three equations are summarized below. The three equations can then be solved simultaneously for the three constants a, b, and c. However, this is not performed here as it is not required by the Exercise. Usually, the various sums are computed first, the inserted into the three equations directly. This is followed by solving the three equations numerically using the MATLAB command `solve` as explained in Chapter 10 on Solving Equations.

$$\sum_{i=1}^{n} y_i = na + b\sum_{i=1}^{n} x_i + c\sum_{i=1}^{n} x_i^2$$

$$\sum_{i=1}^{n} x_i y_i = a\sum_{i=1}^{n} x_i + b\sum_{i=1}^{n} x_i^2 + c\sum_{i=1}^{n} x_i^3$$

$$\sum_{i=1}^{n} x_i^2 y_i = a\sum_{i=1}^{n} x_i^2 + b\sum_{i=1}^{n} x_i^3 + c\sum_{i=1}^{n} x_i^4$$

Seven Science Articles on Nanotechnology, Chaos Theory, MATLAB, Solving Equations, Golden Ratio, etc.

Seven Science Articles on Nanotechnology, Chaos Theory, MATLAB, Solving Equations, Golden Ratio, etc.

www.ingramcontent.com/pod-product-compliance
Lightning Source LLC
Chambersburg PA
CBHW071413150726
48000CB00001B/292

9798869174017